A RELUCTANT ICON

PURDUE STUDIES IN AERONAUTICS AND ASTRONAUTICS

James R. Hansen, Series Editor

Purdue Studies in Aeronautics and Astronautics builds on Purdue's leadership in aeronautic and astronautic engineering, as well as the historic accomplishments of many of its luminary alums. Works in the series will explore cutting-edge topics in aeronautics and astronautics enterprises, tell unique stories from the history of flight and space travel, and contemplate the future of human space exploration and colonization.

(●)

RECENT BOOKS IN THE SERIES

Dear Neil Armstrong: Letters to the First Man from All Mankind
James R. Hansen (Ed.)

Piercing the Horizon: The Story of Visionary NASA Chief Tom Paine
Sunny Tsiao

Calculated Risk: The Supersonic Life and Times of Gus Grissom
George Leopold

Spacewalker: My Journey in Space and Faith as NASA's Record-Setting Frequent Flyer
Jerry Ross

A RELUCTANT ICON

Letters to Neil Armstrong

JAMES R. HANSEN

Purdue University Press
West Lafayette, Indiana

Copyright 2020 by Purdue University. All rights reserved.
Printed in the United States of America.

Cataloging-in-Publication data is on file with the Library of Congress.

Paperback ISBN: 978-1-55753-969-4
EPUB ISBN: 978-1-55753-970-0
EPDF ISBN: 978-1-55753-971-7

The majority of the letters featured in this volume are from the Neil A. Armstrong papers in the Barron Hilton Flight and Space Exploration Archives, Purdue University Archives and Special Collections. Condolence letters courtesy of Carol Armstrong.

"A great man is always willing to be little."
—Ralph Waldo Emerson

I dedicate this book to those rare men and women of the world who are naturally "little," achieve greatness, and always stay true to what they are. For they are the giants of our humankind.

CONTENTS

	PREFACE	IX
1	RELIGION AND BELIEF	1
2	ANGER, DISAPPOINTMENT, AND DISILLUSIONMENT	36
3	QUACKS, CONSPIRACY THEORISTS, AND UFOLOGISTS	55
4	FELLOW ASTRONAUTS AND THE WORLD OF FLIGHT	107
5	THE CORPORATE WORLD	190
6	CELEBRITIES, STARS, AND NOTABLES	214
7	LETTERS FROM A GRIEVING WORLD	272
	NOTES	377
	ABOUT THE EDITOR	383

PREFACE

Why do I want to publish yet another book about Neil Armstrong? It is a fair question and one that I have already been asked by several colleagues and friends while preparing this second book of letters to Neil Armstrong for Purdue University Press. It is likely a question to be asked again. So let me explain.

In 2005 I published *First Man: The Life of Neil A. Armstrong*, intending the book to stand for a long time as the definitive account of Armstrong's life. *First Man* seemed definitive enough: 770 pages, including 64 pages of endnotes and a 20-page bibliography, based on fifty-five hours of exclusive one-on-one personal interviews with Armstrong in his suburban Cincinnati home. Overall, I conducted oral history interviews with over a hundred different people and corresponded by letter, email, and telephone with several dozen more. Along the way I learned everything I could about Neil—from family members (his sister, June; brother, Dean; wives, Janet and Carol; and sons, Rick and Mark); from numerous schoolmates from grade school, high school, and college; from several of his fellow naval aviators, test pilots, astronauts, and NASA officials; from friends, both casual and close; from his associates during his post-NASA years at the University of Cincinnati and those after he entered corporate business; and from miscellaneous others whose lives intersected with Neil's. Furthermore, Neil himself had done something that made *First Man* rather definitive—he authorized it.

PREFACE

When Neil died at age eighty-two on August 25, 2012, I wrote a new preface for *First Man*, one that addressed his death and added a few of my thoughts on the meaning of his life as I understood it as a historian, a biographer, and someone who had gotten to know him rather well. The publisher, Simon and Schuster, placed the new preface at the front of the original 2005 book and issued it as a second edition. However, not until the opportunity came along to publish a third edition in the summer of 2018, in conjunction with the premiere of the Damien Chazelle–directed film *First Man*, adapted from my book, did I have the chance to extend the biography to cover the last seven years of Neil's life, from 2005 to 2012, including a lengthy discussion of his death and legacy.

One might think that, at that point, I would have judged my work on Armstrong to be truly definitive. But I did not. Having thought, written, lectured, and conversed about Neil's life for seventeen years, since beginning my research for *First Man* in 2001, I had not tired of learning about him, asking new questions about his life, hearing new stories from people who knew him, and finding new source materials. "Definitive," I found, was relative. There was more to know, learn, and discover about Neil Armstrong, just as there always is about historical subjects.

The biggest gap in my knowledge about Armstrong derived from the fact that I had not had significant access to Neil's *correspondence* while researching *First Man*. To be sure, I had far greater access to Neil's papers than any other historian ever had. Virtually no scholar had ever had any entry to his private collection of papers, except for perhaps a few items here and there that Neil might have shared with a space historian or two over the years. Still, my access was itself quite limited. Neil had not given me direct access to his files, stored as they were in cabinets within his home as well as in rented storage units in commercial buildings in Lebanon, Ohio, just north of Cincinnati, where Neil had lived with his family on a farm since leaving NASA in 1971. (He would live on that farm by himself starting in 1990, when his first wife, Janet, separated from him and moved to their vacation home in Utah, until 1994 when Neil and Janet divorced. Later that same year he married Carol Held Knight, after which he and Carol built a home in the Cincinnati suburb of Indian Hill on the site of the home where Carol had lived with her first husband, who had been killed a few years earlier in a private airplane accident in Florida.)

PREFACE

Naturally, I would have preferred that Neil give me carte blanche to go freely into his papers as I saw fit rather than to wait for him to show me what he had selected for me to see after I asked him questions about specific subject matter. But Neil would never have gone for that, though I did ask him for that freedom, saying to him that if he would "let me do my thing" he would not have to be driving—multiple times—the forty-eight-mile round-trip up Interstate 71 from Indian Hill to downtown Lebanon to search through dozens of dusty boxes until he found precisely the files he felt he needed to answer my questions. (Typically Neil made those trips in the days right before I would arrive in Cincinnati for a round of interviews with him, heading into his files in response to written questions I had emailed to him some two to three weeks earlier.) Needless to say, he did not always find exactly what he was looking for in his boxes, and I was not always satisfied with what I got to see. Neil was one heck of an engineer, but he was not a historian, and I often wondered what nuggets remained in those boxes that Neil ignored, passed over, or overlooked, him not knowing that something he might regard as trivial, insignificant, or meaningless could have been, from my training and perspective, wonderfully insightful and important.

In the years following the original publication of *First Man*, Neil bequeathed his papers to the archives at Purdue University, his alma mater. But the process of actually getting them to West Lafayette had only started when he died in August 2012. The task of getting them there fell to Carol Armstrong, his widow. As her grief for the loss of her husband was extremely deep and profound, it was many months before Carol was able to go through Neil's things and get the appropriate materials to Purdue. The first time I saw his collection of papers was when I first visited the Purdue archives in the summer of 2015. For the next three years I spent a good part of my summer in the archives, where, for the first time, I had complete access to Neil's papers.

It was his correspondence—the tens of thousands of personal letters, most of them fan mail written to Neil from men and women, boys and girls, of all ages, from all around the world, along with several thousand of his replies, nearly all of them from the years following Apollo 11—that most fascinated me. For as interested as I was in learning more details of his biography, it was the iconography and myth surrounding Armstrong—that is, the different meanings that society and culture over the years had projected onto him as a global icon and symbol, not just of America but of all humankind—that became the major focus of my research.

Formally, the Neil A. Armstrong papers became part of the Barron Hilton Flight and Space Exploration Archives, which had been established in 2011, a year before Neil's death, with generous support from Mr. Barron Hilton and the Conrad N. Hilton Foundation, based on a gift of $2 million. With that money, Purdue created a special collection within its Archives and Special Collections for "the papers of individuals such as pilots, astronauts, engineers, researchers, and others," especially those with Purdue connections that could offer "original, rare, and unique materials" related to the history of flight and space exploration.[1]

The focus on the history of flight at Purdue is not new. It dates back to 1940 when the university library received a gift of aviator Amelia Earhart's papers from her husband, George Palmer Putnam (1887–1950). From 1935 until her mysterious disappearance over the Pacific Ocean in July 1937, Earhart had served as a Purdue career counselor and adviser to the campus's Department of Aeronautics. Although assorted Earhart papers would ultimately come to rest in a few other archives (including the National Archives and Harvard University), the collection at Purdue stands as the largest compilation of Earhart-related papers, memorabilia, and artifacts anywhere in the world. Building on that base, Purdue in the decades following the Earhart gift continued to grow its history of flight collections. But no acquisition compared in significance to the arrival of the papers belonging to Armstrong.

At the time Neil announced he was giving his papers to Purdue, so too did fellow Apollo astronaut and long-time friend and fellow Purdue graduate Eugene A. Cernan (1934–2017), the last man on the Moon. Along with Armstrong, Cernan had cochaired two of the largest fundraising campaigns in Purdue history, totaling nearly two billion dollars by the early 2000s. Soon to follow suit were a number of other Purdue air and space notables, including several of the twenty-three (and counting) astronauts that Purdue has produced over the years. Into the Flight and Space Exploration Archives went the papers of Roy D. Bridges Jr. (b. 1943), who piloted the Spacelab-2 mission (STS-51F) in 1985 and later became director of NASA's Kennedy Space Center and Langley Research Center; Mark N. Brown (b. 1951), who served as a missions specialist on STS-28 in 1989 and was also a crew member on STS-48 in 1991; Mary L. Cleave (b. 1947), who flew two shuttle missions (STS-61-B and

PREFACE

STS-30) and served from 2004 to 2007 as NASA associate administrator for the Science Mission Directorate; Dave Leestma (b. 1949), a veteran of three shuttle flights from 1984 to 1992, logging over 532 hours in space; William R. Pogue (1930–2014), who served as a member of support crews for Apollos 7, 11, and 14 and was the pilot of Skylab 4, the third and final visit to the Skylab orbital workshop, during which he stayed in space for over eighty-four days from November 1973 to February 1974, the longest crewed flight to that date; Kenneth S. Reightler Jr. (b. 1951), who piloted STS-48 in 1991 and STS-60 in 1994, the latter being the first joint U.S./Russian shuttle mission; Jerry L. Ross (b. 1948), who set records with seven shuttle missions and nine spacewalks from 1985 to 2002; Pierre J. Thuot (b. 1955), a veteran of three shuttle missions who spent more than 650 hours in space, including three spacewalks; Janice E. Voss (1956–2012), who set a record for female astronauts with five space spaceflights; and Donald E. Williams (1942–2016), who piloted STS-51D *Discovery* in 1985 and STS-34 *Atlantis* in 1989.

These diverse collections, and others donated both before and after Neil's bequest, will keep historians and other researchers busy in the Barron Hilton Archives for decades to come. But Armstrong's collection will always be the star attraction, for "Neil is there" in the more than 450 boxes of documents he left to his university—in his reports, coursework, research notes, working papers and subject files, notebooks, and training materials; in his scrapbooks, log books, writings, speeches, photographs, drawings, blueprints, and newspaper clippings; and, most vibrantly, in his correspondence.

Samples from Neil's correspondence, particularly his fan mail, provide the essence of *Dear Neil Armstrong: Letters to the First Man on the Moon from All Mankind*, published in October 2019, as well as this new book, *A Reluctant Icon: Letters to Neil Armstrong*. The contents of the first book are organized into the following chapters:

First Words: Mostly letters to Neil in which people, some he knew and some he did not, offered their thoughts on what he should say when he first stepped onto the Moon.

Congratulations and Welcome Home: Telegrams, notes, and letters sent to Armstrong and his Apollo 11 crewmates, Buzz Aldrin and Mike Collins, immediately after completion of their successful lunar mission.

PREFACE

The Soviets: A surprising number of well-wishing telegrams and letters sent to Armstrong by people in the Soviet Union and Eastern bloc countries, including a number of cosmonauts and Soviet engineers and scientists.

For All Mankind: Being the first man on the Moon instantly made Neil a global icon, as demonstrated by letters written to him in the years following Apollo 11 from all around the world, many of them from children.

From All America: Although a global icon, Armstrong was a quintessential American hero, one that attracted (mostly) adoring letters from thousands of fellow citizens, young and old.

Reluctantly Famous: Neil did everything he could to stay out of the pubic glare, but, as is clear in letter after letter from voluminous mail, the fame of being the First Man on the Moon was inescapable.

The Principled Citizen: Letters asking Neil for all manner and degree of participation and support in contemporary civic and business affairs of his city and state.

Understood in their entirety, the letters in *Dear Neil Armstrong*—some 350 of them—shed new light not just on Armstrong's life and his personal views and opinions but even more so on how society and culture projected different meanings of their own making onto the man who was first to step onto another world.

The letters in the first book only scratched the surface of the Armstrong iconography. My original idea was to publish a much larger book with twice as many letters, but such a book would have been unwieldy. So, with the guidance of the Purdue University Press, the decision was made to divide my preliminary book in half and publish a second book of letters sometime in the year that followed the publication of the first book. That second book would focus on other major themes in the larger life story of the First Man: religion and belief; anger, disappointment, and disillusionment (expressed by fans and critics); quacks, conspiracy theorists, and UFOlogists; fellow astronauts and the world of flight; the corporate world; celebrities, stars, and notables; and last messages and letters of condolence. As with the first book, I have provided a great deal of context and commentary for the letters, thereby giving readers a better understanding of who wrote the letters, what relationship the letter writer may have had

PREFACE

with Neil, if any, and how Neil responded to the letters, in those cases when he did, or had his secretaries and assistants do so on his behalf. (Also, as with the first book, original spellings and punctuation have been preserved.) It is my hope that taken together the two books will provide a host of fascinating insights into the public and private worlds of the man who willingly took that first giant leap onto lunar soil, but who in doing so perforce stepped, with great reluctance, into the public eye, not just for a few years but for the rest of his life.

As hard as it has was to pare down the letters from the some 75,000 stored in the Purdue archives to the contents of these two books, I cannot promise that someday there won't be a third book of Armstrong letters, because almost every letter to Neil in the Purdue archives, and every reply from him, offers interesting new insights into who he was, and even more so into who *we* were, in terms of what we thought about our hero and what we wanted from him.

What do I hope to accomplish with these books? Foremost, I hope that people around the world, today and in the future, will better understand and appreciate Neil Armstrong not just as a global icon who stepped down off a ladder and onto the Moon but as a flesh-and-blood human being with faults, defects, and limitations just like the rest of us. I also hope that people will move away from the many myths and common misunderstandings plaguing the historical memory of Neil Armstrong—primarily that Neil, in the years after Apollo 11, became an ultra-private, totally closed off, near-reclusive man. The letters in the two books show that Neil was hardly any of those things. On the contrary, he was very engaged in the world around him, though he had his own particular ways and standards of how he would deal with society and culture.

I also hope people will stop from time to time to think *Shame on us*. Shame on us for not showing more consideration toward our celebrities and great public figures. Day in and day out, we just ask way too much of them.

There was nothing in the letters to Neil, or from Neil, that made me change my basic understanding of him as I presented it in *First Man*. What they did was add depth, richness, and resonance to everything I had already come to understand about Neil as a person and as an icon.

Finally, I want to again sincerely thank the following individuals at Purdue University for all they have done to make these two books of letters to Neil a reality: Tracy Grimm, associate head of Archives and

PREFACE

Special Collections and the Barron Hilton Archivist for Flight and Space Exploration; Sammie Morris, Purdue university archivist and head of Archives and Special Collections; Katherine Purple, editorial, design, and production manager at the Purdue University Press; Kelley Kimm, PUP senior production editor; Bryan Shaffer, PUP sales and marketing manager; Chris Brannan, PUP graphic designer; and Justin Race, the director of the Purdue University Press. Without their dedication and hard work, the two books could not have been produced as beautifully as they are.

I owe a very special thanks to Carol Armstrong, Neil's widow. While this second book of letters was in production, Carol sent me a number of the condolence letters she received following Neil's death in August 2012. I had asked her earlier if she might share a few of these letters for publication in this second book. It was a lot to ask, and I understood completely when Carol told me that going back through those letters and cards would just be too painful for her. It is a testimony to her profound strength of character and her love for Neil that she ultimately chose to share a selection of those letters. Naturally, before I could include those letters in this book, I needed to ask permission of those great friends of Neil's and Carol's who had offered their sympathies, prayers, love, and friendship in these cards and letters. Virtually all of them that I contacted granted their permission. In the last chapter of this book, you will see how extraordinarily special these letters are, and how they present such a wonderful closing testimony to how people felt about Neil, especially those who were lucky enough to know him well.

Neil would be embarrassed by all the attention, as he always was. But in the service of his memory, and what he meant to the world, then and now, the record of his life deserves to be as complete as we can possibly make it.

James R. Hansen
Auburn, Alabama
March 2020

1
RELIGION AND BELIEF

An event as epochal as the first human being stepping on to another world was bound to be enveloped in religious projections, interpretations, and symbolism. Every religion of the world, in one way or another, endeavored to fold the mission of Apollo 11 into its metanarrative—into its belief system and holy stories, thereby linking past, present, and future into the eternal cosmic experiences of humankind.

No religion engaged and translated the transcendent significance of the first Moon landing more than Christianity. In days and weeks surrounding the mission, Christian leaders around the world gave voice to the idea that humankind's trip to the Moon was a "pilgrimage," a "spiritual quest," and that at the heart of all flying, all space exploration, was a religious truth. NASA's master rocketeer and builder of the Saturn V, Dr. Wernher von Braun, who had converted to Evangelical Christianity shortly after coming to the United States after World War II, expressed the sentiment in 1969 for the scientific and technical community: "Astronomy and space exploration are teaching us that the good Lord is a much greater Lord, and Master of a greater kingdom. The fact that Christ carried out his mission on Earth does not limit his validity for a greater environment. It could very well be that the Lord would send his Son to other worlds, taking whatever steps are necessary to bring the Truth to His Creation." Pope Paul VI expressed it for the Catholic world, referring to the Moon landing as "the ecstasy of this prophetic day."[2]

A RELUCTANT ICON: LETTERS TO NEIL ARMSTRONG

The morning Apollo 11 launched, Reverend Herman Weber gave voice to it for Neil's mother and father, Viola and Stephen Armstrong, and for all American evangelicals from his pulpit in Wapakoneta's St. Paul United Church of Christ: "As Thou hast guided our astronauts in previous flights, so guide, we pray, Neil, the esteemed son of our proud community, and his partners, Buzz and Michael, and all others who are involved in this righteous Lunar flight in every station." In a speech to his congregation days after the Moon landing, a minister in Iowa wrote a letter to Viola Armstrong in which he posited, "Could the external presence of Neil Armstrong, the courageous leader, be a symbol of the presence within of the strong arm of the Lord? . . . Their place was the Moon, their ship was the Eagle, which landed on a firm rock at a place called Tranquility Base. Could there possibly be a rock of ages which is a base for all tranquility, for all peace?"[3]

Many theologians, Protestant and Catholic, concurred: "Armstrong's boots, grating on the crisp, dry surface of the Moon, have announced a new theological watershed. That earthly sound on an unearthly body will lead to a profound shift in the faith and basic attitudes of Christians and other believers, a fact that gradually will become apparent with coming generations. . . . It will cause an eventual, and inevitable, modification in the way man comprehends the man-God relationship—perhaps the most important keystone in his ego-structure and in his concept of his place in eternity."[4] The theologians preached that God had put Neil Armstrong on the Moon to show God's greatness in a new light; to reveal God's expansive presence; restore "proper balance" in humankind's outlook on life; and make people believe in God even more deeply than before. "Of course, we knew that the astronauts were religious men," preached one Baptist minister. "They had to be religious. We wouldn't have sent atheists to the Moon or even let them into an astronaut program."[5]

A number of astronauts were, in fact, religious men. Shortly after landing on the Moon, Buzz Aldrin, a Presbyterian, conducted a private communion service inside the lunar module. A few astronauts turned more spiritual as a result of their lunar experience. Apollo 15's James Irwin, who walked on the Moon in August 1971, became an evangelical minister. "I felt the power of God as I'd never felt it before," Irwin declared. Apollo 16's Charlie Duke, one of the CapComs for Apollo 11, became active in missionary work, explaining, "I make speeches about walking on the Moon and walking with the Son."

RELIGION AND BELIEF

Not surprising, Neil Armstrong received hundreds and hundreds of letters over the years from people mainly writing to him out of some religious impulse. Many wanted to share their belief in God, or the message of God's love, with him. Others wanted answers to religious and other fundamental questions, thinking someone like Neil Armstrong must have the answers or at least special insights into the nature of man, Christ's suffering, heaven, and the afterlife.

The truth was, Neil didn't. In fact, he was not a religious man in any doctrinal sense at all. It was something that his mother, Viola, a strongly devout evangelical Christian, could never accept about her son. Whenever his mother spoke about religion, Neil would listen politely and in silence, offering some terse comment only if pressed. (Loving his mother and wanting to save her from her son's lack of belief in the Christian faith, young Neil developed a conflict-avoidance strategy, which then became a part of his personality and the way in which he dealt with many other difficult subjects during his adult life.)

That is not to say that Neil did not believe in God. It is clear that by the time he returned from Korea in 1952 he had become a type of deist, a person whose belief in God was founded on reason rather than on revelation, and on an understanding of God's natural laws rather than on the authority of any particular creed or church doctrine.

While working as a test pilot in Southern California in the late 1950s, Armstrong applied at a local Methodist church to lead a Boy Scout troop. Where the form asked for his religious affiliation, Neil wrote "Deist." The confession so perplexed the Methodist minister that he consulted Stanley Butchart, one of Neil's fellow test pilots as well as a member of the congregation. Though uncertain of the principles of deism, Butchart praised Neil as a man of impeccable character whom he would and, during their flying together, did trust with his life. He had never once heard Neil utter a profanity (unlike many Christians he had met), and nor to his knowledge had anyone else. Taking Butchart at his word that Neil would positively influence young Scouts, the minister gave Neil the position.

By Neil's own admission, he did not become aware of deism through any history or philosophy class. By the time he was in high school, his favorite subject was science, under the direction of department head and dean of boys John Grover Crites. Crites came to Blume High School in Wapakoneta in 1944, the same year the Armstrongs moved to town. A man in his early fifties, Crites taught chemistry, physics, and advanced mathematics; he was the type who, according to one of Neil's classmates, "gave the kids [who shared his interests] all the experience and all the knowledge they could absorb."[6] Living into the 1970s, Crites was available for interviews on the eve of the Apollo 11 mission. "Science was [young Neil's] field and his love," Crites reported to journalists. Neil not only kept "a goal in mind," but "he was the type of fellow who always tested out a hunch. He was always seeking an answer to some future question," always on course to find the "right answer." This critical spirit made him "a natural for research." Crites continued: "Neil was the type of boy who never let anyone know that he knew anything. You had to ask to get an answer, but he expressed himself well in written form." Fellow engineers, test pilots, astronauts, space program officials, and other colleagues would concur. "I did not see Neil argue," remembers NASA mission flight director Eugene Kranz. "He had the commander mentality and didn't have to get angry." According to Charles Friedlander, who directed the astronaut support office at Kennedy Space Center, "I saw that in the crew quarters. If something difficult came up, he would listen politely. He'd think about it and talk to me about it later if he had something to say."[7]

Like many journalists covering the space program, CBS's Walter Cronkite also experienced Neil's nonconfrontational—some have even said evasive—style. On CBS's *Face the Nation* on Sunday, August 17, 1969, three weeks after the Apollo 11 splashdown, the issue of Madalyn Murray O'Hair publicly declaring Neil an atheist resurfaced. Cronkite asked, "I don't really know what that has to do with your ability as a test pilot and an astronaut, but since the matter is up, would you like to answer that statement?" To which Neil replied, "I don't know where Mrs. O'Hair gets her information, but she certainly didn't bother to inquire from me nor apparently the agency, but I am certainly not an atheist." Cronkite followed up: "Apparently your [NASA astronaut] application just simply says 'no religious preference.'" As always, Neil registered another answer as honest as it was vague and nondescript. "That's agency nomenclature which means that you don't have an acknowledged identification or association with a particular church group at the time. I did not at that time." At which point, Cronkite dismissed the matter. According to Neil's brother, Dean, Cronkite on another occasion asked Neil if he felt closer to God when he stood on the Moon's surface, to which Neil gave a totally ridiculous non sequitur: "You know, Walter, sometimes a man just wants a good cigar."[8]

In this chapter readers will see how letter writers projected various religious beliefs onto Neil, the other astronauts, and space exploration in general, and how Neil responded to such letters—not by arguing with anyone, but by ignoring their questions or sidestepping the issue of religion altogether.

"WE UNDERSTAND THAT YOU ARE A CHRISTIAN"

August 1, 1970

Dear Mr. Armstrong:

Our family is concerned about whether men should be going to the moon and other planets or not.

We understand that you are a Christian, so probably you know what is said in Psalm 115, verse 16 ("the heaven, even the heavens, are the Lord's; but the earth hath he given to the children of men").

We would really appreciate your views on this verse and it's meaning and it's possible connection with the rightness or wrongness of journeys into space.

Cordially,

Mrs. Stanley L. Moore
Brookfield, Missouri

NOTE: THE ABOVE IS A COPY OF A LETTER WE SENT TO YOUR HOME IN OHIO. WE HAVE RECEIVED NO ANSWER, SO PROBABLY YOU DIDN'T RECEIVE THE LETTER.

"HOPED THAT THE LANDING WOULD HAVE GREAT SPIRITUAL MEANING"

October 14, 1970

Dear Mr. Armstrong:

On the first anniversary of the moon landing, you, Neal Armstrong, first to step on the moon, said you had hoped that the landing would have great spiritual meaning to the entire world (or words to that effect). I personally had hoped that the "state of mankind message" in my "God & Country" booklet would have had the meaning to which you had referred. So for you, Neal Armstrong, I have written another.

Neal Armstrong (Man of God)

Wasn't it Neal
Who flew away
To land on the moon
Another day?

Wasn't it Armstrong
Who lead the command
On their way
To a foreign land?

But when they touched
How did we feel?
Didn't some gasp?
Didn't some reel?

If I changed it
And reversed the field
As a farmland in draught
Short on yield

If I turned the name
(Reversed it, say)
Would the meaning
Grow this day?

At "touchdown"
("Try this for feel")
Didn't the strong arm
Make us all Neal?

Sincerely,

John Calvin Warder
Fremont, Iowa

RELIGION AND BELIEF

"HALLELUJAH!"

December 23, 1970

Dear Sir and Brother in Christ,

We here in this Christian orphanage are so longing to get from you the *statement* that the first word you said while stepping on the moon in July 1969 was hallelujah!

It seems there are only few people who understood your short prayer and the word hallelujah. Here in our orphanage it is only one person, elsewhere 2 persons, who heard it and the technician from Radio Djakarta, the R.R.I, who heard it the short prayer, but understood not exactly.

When David Itaar, son of the parents of this orphanage, called Pelangi, what means Rainbow. Heard it he told his parents and us and we were enthusiastic and happy. One of the other two persons who heard the word hallelujah is a young lad from Kupang who came from Djakarta and told it, but I do not know her personally, even not her name. The other lady is from Ambois and works in the American Embassy in Djakarta and told me that she *read it* but could not find the place again. I pray for God's Spirit in you, and that His will be done.

Miss H. Arkema
Panti Asuhan Pelangi
Abepura
Irian Barat (West Irian)
Indonesia

"THE BENEFITS OF TRUE CHRISTIAN LIVING"

August 27, 1971

Dear Mr. Neil Armstrong,

While working with young precious children this summer in a Bible school program I had an opportunity to present to our 14 children an inspiring story of your life as presented in a small booklet by Norman Vincent Peale. It was a fine example for young children and it brought home so clearly to them the benefits of true Christian living.

Since then my only son Peter has taken ill with leukemia. He has so much promise, spunk and courage. He seems to take everything "thrown at him." He is unaware of his real illness but knows he is ill. He fights back each time and we sincerely believe he will again make it.

My request is simple: a word of encouragement from a fine American who has placed God and his country as top priority. He is a patient at Children's Memorial Hospital, West 3, Fullerton Avenue, Chicago, Illinois.

Sincerely,

Mrs. A. L. Molina
Joliet, Illinois

Neil sent a short encouraging letter to young Peter Molina but his letter expressed no religious views or messages. The booklet by Norman Vincent Peale mentioned by Mrs. Molina may be a reference to the magazine Guideposts, *which Peale founded in 1945. In the February 1970 issue of* Guideposts *there was an article by Neil's mother, Viola Armstrong, "As told to Lorraine Wetzel," entitled "Neil Armstrong's Boyhood Crisis," in which Viola recounts an incident in which Neil, as a sixteen-year-old, saw one of his fellow students die in a plane crash, causing Neil (allegedly) to question whether he should himself keep flying (he had earned his pilot's license on his sixteenth birthday). According to the story, Neil spent most the next few days in his room, praying and reading the Bible, before deciding that God wanted him to continue flying. In my interviews with Armstrong for* First Man, *Neil asserted that this story was false and that it was a projection of his mother, a devoted evangelical Christian, onto him after he experienced the death of the young man (Carl Lange) in a plane crash outside of Wapakoneta, Ohio, on July 26, 1947, shortly before Neil's seventeenth birthday.*

"A FRIEND OF MINE DOESN'T BELIEVE IN GOD"

September 3, 1971

Dear Mr. Armstrong,

I have a problem which I hope you will help me solve. A friend of mine doesn't believe in God. He thinks science has all the answers.

Because he has a great admiration for you and would respect your opinion, I would appreciate it if you could send him a letter indicating your beliefs and how your knowledge of science has affected them. Thank you for helping me bring the peace of God's love into his life.

Sincerely,

Neil Bunker
University of Wisconsin-Whitewater
Whitewater, Wisconsin
Send the letter to: Cliff Anderson, Jr., Oconto Falls, Wi.

No copy of any letter from Neil to Cliff Anderson exists in Purdue University's Neil A. Armstrong papers collection.

"THE GREATEST GIFT THAT WE COULD THINK OF"

September 28, 1971

Dear Colonel Armstrong,

In July 1969 you and your fellow Astronauts achieved what we human beings had never achieved before you by landing on the moon thereby blazing a terrific trail for mankind. In appreciation of your feat members of my Church wrote you a letter on 25th September, 1969 to felicitate the three of you on your achievement. We were so thrilled by your success that if we had a thousand pounds to spend on a gift it would be inadequate to express our joy and congratulations; so all we could do was send each of you a Bible backed up with our prayers. This, financially speaking, was a very small gift but spiritually speaking and in our manner of thinking it was the greatest gift that we could think of. Since that time my Congregation and I have never slackened in our prayer for you and for other brave men who are following in your footsteps; you have always been in our thoughts.

One outcome of this is that my Congregation and I have decided to honour you further by having the names of the three of you associated with the foundation-laying of one of the Churches which have been building of late and to this end we propose that the foundation stone of our new Church in Lagos which is the capital both of the Federation of the Lagos State be layed in your name by a person of your choice.

When we have heard from you we shall let you know the date of the foundation laying and other particulars.

Yours sincerely,

Prophet C. O. B. Ijaola
Founder & General Superintendent
Christ The Savior's Church (Aladura)
Lagos, Nigeria

> *"I would like to express my appreciation for the honor you have bestowed upon us"*

December 20, 1971

Mr. C. O. B. Ijaola
Prophet, Christ The Savior's Church (Aladura)
Lagos, Nigeria

Dear Mr. Ijaola:

Thank you for your very kind letter of September 28. I apologize for my delayed response.

On behalf of all who were connected with the success of the Apollo 11 mission, I would like to express my appreciation for the honor you have bestowed upon us. I have been in touch with my colleagues and it is our hope that someone from the American Embassy in Lagos might be able to represent the Apollo 11 crew at the foundation laying ceremony for your new building.

Please extend to the members of your congregation our sincere best wishes on this important and meaningful occasion in the life of your church.

Sincerely,

Neil A. Armstrong

RELIGION AND BELIEF

"BOW DOWN TO GOD IN PEACE"

December 4, 1971

Dear Mr. Armstrong,

I sent a copy of this poem to your mother, as this was my tribute, the best way I could express it. I am in my seventies and do not write so good. If you want to keep this, have you a copy printed. Your mother wrote that she would keep her copy.

   I saw you on television and decided to send this which I wrote when landed on the moon.

> "To Our Astronauts, July 20, 1969"
>
> There is a time and place for everything 'neath the Sun'
> Our new discoveries have just begun
> To-day on, the moon, a man did land
> I saw him scoop a hand of sand.
> On the moon, I saw him walk
> From a faraway distance I heard him talk.
> The mystery of God, o'er and o'er
> Man has tried many ways to explore.
> On the wings of the Eagle you flew away
> Hoping to come back to us some day.
> You flew into the mysteries of God
> Where, no human foot had ever trod.
> You left footprints in the sand
> In a faraway country you did stand.
> Now, of all these things we can brag
> You planted there our American flag.
> O'er that land, may she proudly wave
> In honor of the names of our brave.
> Armstrong, Alden, and Collins you will see
> Your names go down in history.
> May all the world bow down to God in peace
> To live in harmony, 'til time shall cease.
>   —Edna Sweeney

A RELUCTANT ICON: LETTERS TO NEIL ARMSTRONG

"Thank you for sharing your poem"

January 3, 1972

Mrs. Edna Sweeney
Dayton, Kentucky

Dear Mrs. Sweeney:

Thank you for sharing your poem honoring Apollo 11 with me. The tributes paid the flight of Apollo 11 have been most gratifying and I appreciate your taking the time to send me your reflections about the first moon landing, in which I had the honor to participate.

Sincerely,

Neil A. Armstrong
Professor of Aerospace Engineering

"WE ARE TRULY BLESSED"

December 17, 1971

Dearest Mr. and Mrs. Armstrong:

Although you do not know me, I, like people all over the world, was only one of the millions of people praying for you on your history-making journey to the Moon—and I feel we are truly blessed in this beautiful country of ours to have men like you, with all the courage and devotion which you have. May you and your family be truly blessed in the next year and all the continuing years to come.

    I also wish you well in your next career and feel the young men fortunate to have you as a teacher are ever so lucky.

    May God bless you and yours always.

    Peace for Mankind.

Mrs. Betsy Best
Long Beach, California

Mrs. Best had seen a newspaper story relating that Neil had resigned his NASA position and was becoming a professor of aerospace engineering at the University of Cincinnati.

"PRAY ON SPACE TRIPS"

January 5, 1972

Dear Neil Armstrong,

My name is Brian Stuart. I belong to South Jefferson Christian Church. For a merit badge for Bible Scouts I had to write an astronaut and encourage him to pray on space trips. I hope you will.

Your friend,

Brian
Valley Station, Kentucky

"ASKING FOR YOUR HELP"

August 16, 1972

Dear Sir:

The parishioners of St. Paul's Catholic Church of Jacksonville Beach, Florida are preparing to put on their annual fiesta to be held September 30, 1972.

In an effort to make the most successful Spanish fiesta ever, we are asking for your help. In a particular booth, we are going to raffle off items of a personal nature from prominent people, such as yourself, with proceeds going to the church.

The item can be anything of your choice, preferably something you feel would be a keepsake by a delighted owner.

If you are able to assist us, we would be most grateful.

Please send the item to the following address, if possible no later than September 20, 1972. [Address withheld.]

Sincerely yours,

Nancy McMann
Chairman Raffle
St. Paul's Parish
Jacksonville Beach, Florida

Reply from Fern Lee Pickens, Neil's assistant

September 15, 1972

Mrs. Bruce McMann
Daytona Beach, Florida

Dear Mrs. McMann:

Mr. Armstrong has asked me to thank you for your letter offering him the opportunity to participate in your fund-raising project by contributing a personal memento.

We regret we are unable to honor your request. He receives so many similar requests that, in fairness to all concerned, we have found it necessary to decline all of them.

We do appreciate your interest and hope your project will be a success.

Sincerely,

(Mrs.) Fern Lee Pickens
Office of Public Affairs
NASA Headquarters
Washington, D.C.

"DO YOU PRAY?"

September 25, 1972

Dear Mr. Armstrong,

I am a eight grade student at St. Michael's School, Findlay, Ohio. In our religion class we are studying prayer, and is part of a project. I am asking famous people in the world, their ideas about prayer. Do you pray?

RELIGION AND BELIEF

If you do, would you share your thoughts about the value of prayer? Would you send me a copy of a favorite prayer?

I would like to congratulate you for the lovely moon walk.

Sincerely,

Linda Smoke
St. Michael School
Findlay, Ohio

Reply from Ruta Bankovskis, Neil's secretary

September 26, 1972

Linda Smoke
Saint Michael School
Findlay, Ohio

Dear Linda:

Professor Armstrong has asked me to reply to your letter asking him for his thoughts on the value of prayer. We are unable to honor your request as Professor Armstrong considers this too personal a matter to make public.

He asked me to thank you for your interest.

Sincerely,

Ruta Bankovskis
Secretary to Professor Neil A. Armstrong

"WHAT YOU FELT IN A 'RELIGIOUS' VEIN"

October 11, 1972

Dear Sir:

I am presently a Senior of Aerospace Technology at Arizona State University. I am enrolled in a course of Aerospace Systems Design, and we have begun studies on life support systems. I would like to report on Pressurization, and this will be my secondary theme. However, as we got onto the subject of isolation, we learned; "Nearly every astronaut has returned with a changed psychology, and in general, more religious." This interested me to a great extent and I would enjoy writing a paper on this if I'm allowed to write a non-technical paper. Otherwise I will turn it in as a minor. This brings me to the subject of this letter.

I would deeply appreciate a reply (general or specific, short or long) concerning yourself and what you felt in a "religious" vein (Christianity included) as you viewed earth, space, or at any moment in your flight.

I am enclosing a stamped, self-addressed post card which I would appreciate you dropping in a mailbox if you wish not to reply. This will give me a chance to abort and write on a different subject if necessary. Obviously without "personal opinions" I would be hard-pressed for data.

Thank you for your time.

Sincerely,

Bob Diehl
Tempe, Arizona

"I'M A BORN AGAIN CHRISTIAN"

November 30, 1972

Dear Mr. Armstrong,

I am a great fan of yours. My name is Bob Reader. I am 12 years old. I have 5 sisters but no brothers or dad. I'm a junior astronomy here in Grand Rapids. I've a scope of my own.

I heard that there is to be another launch on December 6. I do pray it will be a very good one.

Could you find it in your heart to write me? I'm a born again Christian and thank God for good men like you. I would concider a great honor "please." Your the kind of man I would like to grow up to be.

Thank you very much.

Your true friend.

Bob Reader
Denison, Michigan

RELIGION AND BELIEF

At the bottom of his letter Bob gave Neil his phone number. The launch referred to in the letter was delayed to December 7, 1972. It was the launch of Apollo 17, the last of the Apollo manned lunar landing missions.

Reply from Fern Lee Pickens

February 20, 1973

Master Bob Reader
Denison, Michigan

Dear Bob:

Mr. Armstrong has asked me to thank you for your letter and the fine compliment you pay him by desiring to grow up to be like him. He is always interested in hearing from young people who are interested in the space program and would like to answer personally all the letters he receives. However, this is not possible because of the demands on his time.

We are enclosing an autographed picture of Mr. Armstrong and some information which we hope you will enjoy.

Sincerely,

(Mrs.) Fern Lee Pickens
Office of Public Affairs
NASA Headquarters
Washington, D.C.

"LITTLE PROCESSION TO THE HIGH ALTAR"

June 21, 1974

Dear Mr. Armstrong:

The time is approaching the service that will be held in the Cathedral to commemorate the fifth anniversary of the Apollo XI mission. The form of that service, about which I wrote you earlier, is now rapidly taking shape.

A RELUCTANT ICON: LETTERS TO NEIL ARMSTRONG

The lunar sample that is to be presented to the Cathedral for inclusion within the beautiful new "Space Window" has been prepared at the Johnson Space Center in Houston, delivered to the NASA here and will shortly be brought to the Cathedral for safekeeping until the service on July 21st.

I know when you wrote me last you were not absolutely certain whether you would be able to accept my invitation to attend. I hope that your plans have developed in a way that will permit you to honor us with your presence. In that case may I ask you to be good enough to make a ceremonial presentation of the lunar material to me in the course of the service. I shall invite Mike Collins and Edwin Aldrin to accompany you in the little procession to the high altar where you would offer the rock with a graceful word which we would be glad to prepare for you.

I very much hope that you may be able to be here and to take such a significant part in a ceremony that will, I think, be deeply meaningful to our whole nation.

Faithfully yours,

The Very Reverend Francis B. Sayre, Jr.
Dean
Washington Cathedral
Mount Saint Alban
Washington, D.C.

"I am wary of how it will be reported in the press"

July 15, 1974

The Very Reverend Francis B. Sayre, Jr.
Washington Cathedral
Mount Saint Alban
Washington, D.C.

Dear Dean Sayre:

I was pleased to find your letter awaiting my return to the university after several weeks' absence. I do not teach during the summer months and visit only occasionally.

I'm pleased to hear that all is well with the plans for the ceremony of

21 July. I do plan to attend.

Although I have no reservations about participation in the service in the manner suggested, I am wary of how it will be reported in the press. Most newspapers will say that the Apollo 11 crew gave a lunar rock to the Washington Cathedral.

The readers' impression that the sample was previously in the crew's possession will be misleading. Of course, the crew could represent NASA, but inasmuch as none of us are now affiliated with the agency, the proper impression will be difficult to create.

I personally believe that the most straightforward solution would be to have Dr. [James] Fletcher [the NASA Administrator] make the presentation. I know you will consider these thoughts as you formulate your final program.

Sincerely,

Neil A. Armstrong
Professor of Aerospace Engineering

"Living symbols of the great leap"

July 26, 1974

Dear Mr. Armstrong:

In behalf of Dean Sayre may I tell you what a great honor it was to have you participate in the moon rock service here in the cathedral last Sunday. You and your colleagues certainly are under great burdens as living symbols of the great leap; I suspect there must be times when that burden is light and others when it is annoyingly heavy. I hope that Sunday was one of the former!

Thank you for helping to make that day one of the most historical in the annals of the cathedral.

Faithfully yours,

Jeffrey P. Cave
Canon Precentor
Washington Cathedral
Mount Saint Alban
Washington, D.C.

Neil's Apollo 11 crewmates, Buzz Aldrin and Michael Collins, also attended the ceremony at the Washington National Cathedral on Sunday, July 21, 1974. Together they presented a fragment of a Moon rock across the cathedral's Jerusalem Altar to Reverend Sayre, who then held the rock aloft before the crucifix. The fragment was 7.18 grams of basalt lunar rock estimated to be some 3.6 billion years old. It was embedded into the Scientists and Technicians Window on the south side of the cathedral, in a nitrogen-filled capsule sitting at the center of a section of red glass representing the planet Mars. The beautiful stained-glass creation, which became familiarly known as the Space Window, was designed by St. Louis artist Rodney Winfield to commemorate America's exploration of space and man's first steps on the moon. Winfield's intention for the window was "to show the minuteness of humanity in God's universe."[9] Besides the representation of the Red Planet, the artwork also features whirling stars and orbiting planets, featuring colors inspired by NASA photographs taken during the Apollo 11 mission. Divine rays of light radiate from numerous solar spheres, and white dots of stars shine through the dark background. There are also thin silvery trails depicting the path of the Apollo 11 spacecraft. As tourist guidebook Atlas Obscura described it, the Space Window is the embodiment of "the intersection of religious thought with the spirit of exploration and the mysteries of the universe"[10]—in sum, the glorious meeting of science and religion.

As Neil suspected, stories about the July 21, 1974, service, universally reported that the Moon rock was "a gift from the crew of Apollo 11."

Significantly, the Washington National Cathedral, on September 13, 2012, was the scene of the large public memorial service honoring Neil's life following his death on August 25 from complications following cardiac bypass surgery in a suburban Cincinnati hospital. Before an overflowing crowd, Mike Collins led the mourners in a prayer. Eulogizing Neil was his good friend and Purdue mate Gene Cernan, the Apollo 17 mission commander and last man to walk on the Moon, and Charles Bolden, the NASA administrator. Also speaking at the service were John H. Dalton, former U.S. Secretary of the Navy, and the John W. Snow, a fellow Buckeye and former CEO of CSX Corporation who had served as U.S. Secretary of the Treasury under President George W. Bush. One of Neil's favorite contemporary singers, jazz contralto Diana Krall, sang Fly Me to the Moon. Although Neil was a deist and not a doctrinally religious man—a fact not acknowledged during the church service, of course—the Reverend Gina Gilland Campbell read a passage from the Book of Matthew, and the Right Reverend Mariann Edgar Budde delivered a homily.

RELIGION AND BELIEF

"THE WORD OF GOD MAKES IT PLAIN"

July 17, 1974

Dear Neil,

As I read the article in the *U.S. News & World Report* of you stating "Where one is going?" my heart was touched that *you do not know.* The word of God makes it plain. Please read the booklet "No Detours to Heaven." Only two ways where man can go. He makes his choice.

Sincerely,

Mrs. Gladys B. Crocker
Tempe, Arizona

Neil's question cited in the U.S. News and World Report *story concerned the direction to be taken by the U.S. space program, not to anything related to humankind's spiritual life.*

"YOUR AMERICAN AND CHRISTIAN STAND"

November 20, 1974

Dear Dr. Armstrong:

This is a request for a big favor—for the many teenagers who need to sense anew your American and Christian stand.

Since I can't interview you in person, I've depended on Ohio friends who have talked some with your Mother, and numerous articles I've collected.

I'm a free-lance writer of Christian juveniles. I enclose a copy of a story I'd like to send to a publisher who has already accepted other features.

Will you please correct where necessary, then write a statement for the publisher giving me permission to submit this story?

I will be deeply grateful for this courtesy. I do not want to send it without your knowledge.

In sincere appreciation,

A RELUCTANT ICON: LETTERS TO NEIL ARMSTRONG

Laura S. Emerson
Assoc. Prof. of Speech Emeritus
Member of National League of American Pen Women
Marion, Indiana

For many years Professor Laura S. Emerson taught speech at Indiana Wesleyan University, a private Christian liberal arts university located in Marion, Indiana, which is an institution of the Wesleyan Church, an evangelical Protestant denomination. In her letter Dr. Emerson was most likely making reference to the story in Guideposts *that Neil's mother related about his "boyhood crisis."*

"Your sources are inaccurate"

November 22, 1974

Professor Laura S. Emerson
Marion, Indiana

Dear Professor Emerson:

I very much appreciate your courtesy in sending your draft to me for approval.

I must be candid and tell you that I do not wish stories of the type you have prepared to be published. If I did, I would certainly respond more favorably to the many inquiries I receive from editors and publishers; and write them myself.

Your sources are inaccurate. People remember and report what they wish to remember. In my opinion, the anecdotes bear little resemblance to the facts as they occurred.

I very much hope that you will accept my request to refrain from publication.

Sincerely,

Neil A. Armstrong
Professor of Aerospace Engineering

In typical fashion, Neil declined to provide any such statement to the college professor—nor would he ever comment publicly to anyone else on his religious beliefs. Privately, as stated in an earlier notation in this chapter, Neil was a deist, a person who accepted the rational form of spiritual belief that grew out of the eighteenth-century Enlightenment and Age of Reason, which posited God's existence as the "master architect" and cause of all things but which rejected divine revelation and direct intervention of God in the universe by miracles. Of course, Neil shared none of that with Professor Emerson—or anyone else.

"A CLASSMATE OF MINE FROM THE SEMINARY"

July 27, 1975

Dear Mr. Armstrong—

Happy and prayerful greetings to you and your family. I hope you are having a most pleasant summer.

I am really most reluctant to disturb you like this, but on the chance it wouldn't be too much of an inconvenience, may I ask a favor of you. A classmate of mine from the seminary—and a fellow missionary in The Orient (Taiwan & Hong Kong)—will be visiting me in Cincinnati next week (Aug. 4–8). He is a real "space-buff" and was one of the most excited and thrilled of the people on this globe who watched your moon-landing. I know it would be the Thrill of his life for him if we could get a chance to meet you, just to say "hello," for a couple minutes. He's on his way back to Taiwan after a six week home leave, and this would be a great send off for him.

I could arrange to be at any place and at any time if this is convenient for you. If not, I completely understand, and I'm really sorry to have taken up so much of your time already.

Just in case, I'll take this opportunity to say thanks for the Thrill of a lifetime for all of us. After six years, it's still hard to believe it.

Sincerely,

James Huvane
The Maryknoll Fathers
Cincinnati, Ohio

P.S. I'll enclose a self-addressed envelope to facilitate this for you.

Apparently Neil agreed to meet with the two missionaries, as there is a cursive notation in the hand of one of Neil's secretaries at the top of the letter: "Meeting cancelled by Rev. Huvane due to funeral."

"WE ARE ALL ON REMOTE CONTROL FROM GOD"

December 3, 1978

Dear Neil,

Well so far I cannot get anyone to believe me that I met the spirit of Dad on train wheels in my head then three months after you landed on the moon I brought Dads screaming soul out of the cemetery, and put it up into the heavens. I believe my son is God and his sister the Blessed Virgin. I suppose only time will tell then in the year 2,000 when the skys open up than we will all be living back in the mountains. I cannot wait for that day.

Do you remember telling me you were going to be the first man to land on the moon and I said nothing would be worry me because I would be living with God and his mother it seems as though both of our dreams have come true we were only children ourselves when we said that.

After I went home out of hospital after I had my son my head went on to further train wheels and I was completely knocked out to it, it was a pleasure to open my eyes and know that I was alright. I feel as though I have experienced death.

There is nothing you or anyone else can do for me except that I want you to know that I am waiting for you to come and take us all back to the mountains. I hope you and your family are well. I read about you losing your finger I hope you are now alright. God did that to you, seen we are all on remote control from God.

I remain

Mrs. Jeanette Lorraine Dent
Brisbane
Queensland, Australia

RELIGION AND BELIEF

"WE WOULD ALL SEEK THE REAL PERFECTION"

August 20, 1979

Dear Mr. Neil Armstrong,

Here are some Christian thoughts that may interest you, I hope.

It would be nice if everybody wouldn't consider everybody else as a sex symbol first, then an ego symbol, because that way birds of a feather fly together. And while I like birds, I think human beings have much more of a task in life than groups flocking and flying together.

If everybody claims to be a human being, Mr. Neil Armstrong, let them prove it to everybody who is not a killer, and to everybody who is not a trouble looking, or trouble seeking, predator, Mr. Neil Armstrong, predator like other birds of a feather like to flock together, no matter what their religion, nationality or race. They range from mental predators, to physical predators to sexual predators, they all have one thing in common, they dislike and some even despise, those persons or people they consider weak. Some of them even despise babies, because babies shouldn't be weak either. These mental, physical, and sexual predators who despise the weak, Mr. Neil Armstrong, what makes them think they are really brave and strong? A couple of good punches would flatten them for life, for they only have the courage to fight those they consider a lot weaker than they are.

Mr. Neil Armstrong, I think God gives to some persons who may be dumb, guilt feelings, so that they will know when they are doing something wrong. I think God gives them also awareness of those dangers, and people dangers around them, for their own satisfaction. Mr. Neil Armstrong, I've often seen babies with nipples in their mouths and I've wondered does a nipple in the mouth of a baby give it mental security. Why would a baby seek mental security in a nipple, when he or she has a mother? It must be really terrifying to some babies, because they have no way of protecting themselves.

[This letter continues for another five pages, with nine more references to "Mr. Neil Armstrong." The letter then concludes as follows.]

Mr. Neil Armstrong, with God's government of humanity and its truths in our minds daily, and its rights and wrongs in our minds daily, and its justice in our minds daily, who would want to stay here at all?

Mr. Neil Armstrong, with all of our minds that perfect in regard to everybody else, we would all seek the real perfection: God's mind without our bodies.

Sincerely,

[Name withheld]
Cincinnati, Ohio

Neil instructed his secretary not to respond to this letter.

Over the years, Neil received hundreds of very curious cards and letters, many of them from crackpots whose words make no sense but also from many others, though undoubtedly eccentric and offbeat, who offered (mostly) sensible thoughts. Neil almost never answered any such letters himself; if the letter writer received any reply, it was a form letter written and signed by Neil's secretary. In this case, he instructed his secretary, Vivian White, not to respond to the letter.

"DO YOU BELIEVE IN TRANSMIGRATION OR REINCARNATION?"

April 4, 1980

Dear Mr. Armstrong,

Your hosting of the PBS-TV series on Charles Darwin leads me to believe you fully accept Darwin's theory of evolution.

Yet his "survival of the fittest" idea seems to be in error, inasmuch as the unfit in Africa, India, & Latin America seem to survive despite inadequate diets.

However, the Bible and other "holy books" appear to be only various theories about man's alleged "soul" and "spiritual nature." Since they all differ, they can hardly be said to be "divinely inspired" or "revealed" or "dictated by God." Even if one accepts these as allegories, symbols, parables, etc., much they have to say are folklore (such as Samson killing off thousands of Philistines with the "jawbone of an ass"), or obviously untrue (as Joshua's commanding the Sun to stand still), or inconsistent (as Jesus being taken to the cross at 9:00 A.M. in one gospel and at 12:00 noon in another).

Man appears to be an aggressive animal—as the Europeans and British taking the land from the Indians, or the Jews reclaiming their "holy land" after being gone 1,000 years and using the Old Testament "prophecy" as an excuse for their aggressive activities.

But the complexity of organisms—the grasshopper, the bird, man's body and brain—make one wonder how it could have evolved to such near perfection after even thousands or millions of years.

Man does not seem to have free will, as claimed by the Jews, the Catholics, the Methodists, and the Hindus. Even Adam and Eve could not restrain themselves from eating the "forbidden fruit" after being made in "God's image."

Aristotle considered the "soul" as simply the life principle and, unlike Paul, gave the carrot, tree, bird, fish, and dog a soul as well as man. Yet Aristotle believed some part of man's intellect survived death, and seems not to have believed in evolution.

Do you believe in transmigration or reincarnation?

Do you believe in astrology? For example, the solar eclipse of July 20, 1963, fell on the natal Saturn of John F. Kennedy and four months later (the usual time for an eclipse to take effect) he was assassinated. If the eclipse was the "cause" of his death, we could say it was caused either by the eclipse and perhaps due to some misdeed in his previous life; or we could say it was simply a statistical "correlation," the "cause" being unknown or uncertain (as, a number of children have lumps around their necks below the jaws, and so the doctors say they have mumps).

In the book of Job we have a mystery story without a solution, for if Job being a very good man nevertheless suffered much misfortune, it hardly makes sense, since logically good people go to heaven and enjoy life on earth as well and bad people suffer and go to hell, robbers go to prison and non-robbers go free. For Job to say he accepted his misfortune and the wisdom from an inscrutable God hardly does more than beg the question. The same reasoning applies to the story of Jesus and the blind man.

I would like to ask your opinions and if you have books on the subject to recommend which would shed further light. I would be glad to read them.

Sincerely yours,

Buell D. Huggins
Herrin, Illinois

A RELUCTANT ICON: LETTERS TO NEIL ARMSTRONG

Reply from Vivian White, Neil's assistant

April 21, 1980

Mr. B. D. Huggins
Herrin, Illinois

Dear Mr. Huggins:

Mr. Armstrong acted as the host for the BBC production, "The Voyage of Charles Darwin." He took no position on the controversy or the theological implications. He considers that many others are better qualified.

Thank you for your interest.

Sincerely,

Vivian White
Assistant to Neil A. Armstrong

If Neil had wanted his assistant Vivian White to provide a more direct answer to Mr. Huggins's questions, her reply would have made it clear that Neil was a strong believer in modern science, including the theory of evolution. The letter from Mr. Huggins was one of several letters that Neil received—most of them negative—about Neil hosting the series about Charles Darwin.

"BIBLE PROPHECY WILL BE FULFILLED"

August 5, 1982

Dear Mr. Armstrong:

Here's wishing you a happy and healthy 50th.

The last time I heard from you was 13 years ago (as per enclosed copy of your letter.)

I have read that you will be appearing at the Youth Foundation International Leadership Seminar at the Conrad Hilton Hotel in Chicago and would like to meet with you to discuss very important issues.

Two years ago, I contacted Mrs. Gertrude Smith of your City Manager's office, who in turn got your secretary to call me and who, in turn, refused to permit me to speak to you.

RELIGION AND BELIEF

You were destined to be the first man on the moon before you were born and you still have another mission to perform.

Incidentally, at the same time that you were registering at Purdue in July 1953, I was at the Pentagon with Maurice B. Graney, Dean of the College of Engineering at Purdue.

There are numerous other coincidences which will be told and documented at our meeting.

Whether we ever meet or not, Bible prophecy will be fulfilled. I can be reached at the above address or phone.

Yours in Christ,

H.J. Macie
Bulk-Pal
LA GRANGE, ILLINOIS 60525
[Phone number withheld]

On this letter Neil wrote, "No reply." In other words, he did not want his secretary to respond to Mr. Macie with any sort of card or letter.

August 5, 1982, was Neil's fifty-second birthday.

"YOUR CONVERSION TO ISLAM"

September 10, 1982

Dear Mr. Armstrong,

Asalam-alaikum. I wish to congratulate you on the wonderful news of your recent visit to Egypt and your conversion to Islam. The story that is being told in our mosques and in our local newspapers is that after registering in a hotel in the city center of Cairo, you walked with a limp to your room where you rested after your weary journey from the U.S. As you lay in bed, suddenly you heard our call to prayer, *Allah akbar Allah akbar!* Hearing this call, you realized it was not the first time you had heard this cry. You had also heard it while walking on the Moon! But not knowing then what it was, now in Cairo you walked down to the hotel desk and asked about what you were hearing. The hotel clerk told you, "It is our call to prayer, the call to all Muslims to go to the mosque for prayers performed five times a day." I have heard that you then thought to yourself, "O Allah the Blessed, O God, now I remember that same calling was there for me on the surface of the Moon! I heard it there for the first time in my life, and now here again in Egypt I hear it on the earth." A few months later, you expressed in an interview that you had converted to Islam. The call from Allah to you was true.

My brother in Islam, please declare your submission to Allah in the strongest terms! You will be welcome at mosques throughout the Islamic world. Do not let the American government stop you from declaring your faith!

La hawla wala quwata illah billah (There is no strength nor power except Allah).

Fi Amanullah (May Allah protect you).

Mualaf Subhanallah
Cairo, Egypt

In the five decades since Apollo 11, stories have circulated all around the Muslim world that Neil Armstrong converted to Islam. Typically the stories begin with the assertion that when Neil was on his Moonwalk, he heard a voice singing in a strange language that he did not understand. Only later, after returning to Earth, did Neil realize that what he heard on the lunar surface was the adhan, the Muslim call to prayer. He then allegedly converted to Islam and moved to Lebanon—the country in the Middle East, not Lebanon, Ohio, where he actually moved. Many of the stories also relate that Neil subsequently visited several Muslim holy places, including the Turkish masjid where Malcolm X once prayed. By the early 1980s, rumors of Neil's conversion to Islam had grown so far and wide that Neil found it necessary to respond—and to get some help in doing so. In March 1983, the U.S. State Department sent the following message to all embassies and consulates in the Islamic world:

1. Former astronaut Neil Armstrong, now in private life, has been the subject of press reports in Egypt, Malaysia and Indonesia (and perhaps elsewhere) alleging his conversion to Islam during his landing on the Moon in 1969. As a result of such reports, Armstrong has received communications from individuals and religious organizations, and a feeler from at least one government, about his possible participation in Islamic activities.

2. While stressing his strong desire not to offend anyone or show disrespect for any religion, Armstrong has advised department that reports of his conversion to Islam are inaccurate.
3. If post receive queries on this matter, Armstrong requests that they politely but firmly inform querying party that he has not converted to Islam and has no current plans or desire to travel overseas to participate in Islamic religious activities.

The State Department's message seems mostly to have provoked the situation, with many of the Muslim faith suspecting that the U.S. government was forcing Neil to hide the truth, not wanting its great American hero to be known as a Muslim, and thus compelling him to deny publicly his actual religious beliefs, which it was well known Neil would not discuss. Requests for him to visit Muslim countries and attend Islamic events became so frequent by the mid-1980s that Neil set up a telephone press conference to Cairo, Egypt, where a substantial number of journalists from the Middle East were told directly by Neil that there was no truth to the rumor. But his denials did not matter. Once when he was visiting his old fraternity house at Purdue, he was approached by a student of Middle Eastern descent. The young man's father, a Purdue professor, had told him about Neil's conversion to Islam. The student asked Neil whether it was true and Neil explained that it was not. The young man did not believe Neil and told him so. He had been convinced by his father that Neil would lie about it.

"THE 7TH TRUMPET OF REVELATION SHOULD BE SOUNDED"

June 1, 1983

To Mr. Neil Armstrong:

Dearly Beloved,

Greetings and peace love and all joy!
    I must in sincerity believe strongly that the Day of the End is exceedingly near. I would propose to you that the Seventh Trumpet will be sounded on 7/31/83, God willing. I will herewith endeavor to explain my reasoning to you:

A RELUCTANT ICON: LETTERS TO NEIL ARMSTRONG

1. I sought permission from my natural earthly (*Father*) that I should be revealed before the American Apollo 8 Astronauts when they circled the moon on my birthday 12/24/68. He said yes and I the (*Son*) was revealed in accordance with Biblical Prophecy. At that time my Father decided that the 7th Trumpet of Revelation should be sounded on my sister's birthday (which sister is an emblematic incarnation of the *Holy Spirit*), which birthday is 7/31, exactly 7 months and 7 days after my own birthday. (The purport of this missive is largely to illustrate the cooperative conjunction of numerical meanings themselves with the Word, thus hoping to justify the date I give for the Trumpet, *7/31/83*.)

2. The date 7/31/83 is exactly 14 (7 plus 7) years 7 months and 7 days from the Apollo sighting of myself and the Family of God on 12/24/68, and of course July is the 7th month, a fitting time for the 7th Trumpet.

3. Apollo 11 landed on the moon on 7/20/69 and splashed down on 7/24/69. Before Neil Armstrong stepped out of the spacecraft the men of Apollo witnessed the Angelic Space Vehicles (UFO's) that I sent to greet them, (they photographed these). From 7/20/69 to 7/31/83 it is 14 (7 plus 7) years and 11 days—compare (Apollo 11). From 7/24/69 to 7/31/83 it is 14 years and 7 days. From the time Apollo 8 sighted me beyond the moon on 12/24/68, until the time Apollo 11 returned, it was exactly 7 months. Apollo 11 made *31* revolutions around the moon. My sister, Margaret Alyss Thomas Purdy, the incarnation of the Holy Spirit, was born on *7/31/31*. Apollo 8 made *10* revolutions around the moon. The date I am seeking for the Seventh Trumpet, in fact, from the original coming of the Holy Spirit to the Church, to the birthdate of the incarnation of the Holy Spirit which, God willing, will be the date of the 7th Trumpet. My sister, and mother (who is the Virgin Mary), and I were all baptized on 7/24/42. (Splashdown for Apollo 11 – same date) I am now 43 years old, I was baptized in the year 42 and it is 41 years and 7 days from Baptism to Trumpet.

4. My birthdate is 12/24/39, my consort's (Elaine's) birthdate is 4/7/43. In 1983 before her birthday I was 43 years old and she was 39 (corresponding reciprocally to the years of our birth). Now (in June 83) she is 40 and I am 43 (40 plus 43 equals 83, corresponding to the year 83.) (Halfway between my birthday and hers is Valentine's Day 2/14. Halfway between her birthday and mine is the Feast of the Assumption 8/15, the principal feast of the Virgin Mary, my mother Millicent Edna

RELIGION AND BELIEF

Recker Thomas.) Her (Elaine's) birthday is a central date for the celebration of Easter and my birthday is Christmas. Our first Easter in the White House will be 4/7/85. (My retirement began (as originally calculated) on 7/7/77.)

5. From Orthodox Easter 1981 (The first Easter of Reagan's Presidency (who is the Beast of Revelation's 7th Head)) to election day 1984, 11/6, when I will be elected President, is 1,290 days as per Daniel 12:11. (The date of Orthodox Easter in '81 was 4/26.) Add another *45* days (See Daniel 12:12), and you have 12/21, usually the first day of winter, the day when days start getting longer, and only 3 days from my 45th birthday. (Symbolically, the Son (Sun) is buried on the (Solstice) Black Friday and is raised on the Third Day Easter (My Birthday) Christmas.) So if you're willing to wait (as per Daniel) you will be blessed, in the feast of Victory!

6. 1983 is the 500th anniversary of the birth of Martin Luther and I was baptized and confirmed a Lutheran as was my sister, and several of the Apostles. And since it is thought to be the 1950th anniversary of the Resurrection, the Pope John Paul II has declared it a Holy Year for the Catholic Faith. (And it is the 100th anniversary of my Alma Mater the University of Northern Iowa—which holds many rich associations for my family and Elaine and I and the Apostles.)

7. Elaine's and my daughter Eda (after my mother's middle name—the Virgin Mary) and Alyssia (after my sister's middle name—the Holy Spirit) was born on 9/8/67, the traditional birth date of the Virgin Mary. My daughter Joanie Sue (by my ex-wife Lydia Banks Thomas Geoffrey), was born on 6/24/68, exactly 6 months from my birthday, on the Feast Day of John the Baptizer. She was confirmed in the Lutheran Church (Family of God) on Pentecost Sunday '83. My son Jan Ellison was born 3/7/72, the Feast Day of St. Thomas Aquinas. He is 11 now and in '83 begins to play in a baseball league at Apollo (11?) Park. (There is considerably more to tell about the symbolisms of the Apollo flights, mythologically, etymologically, and so forth, yet this will have to suffice for now.)

Therefore be alerted and aroused to readiness for I must believe the time is at hand, God willing. You will hear much more from me when Elaine (Consort Sophia Sakti Radha—Uma) comes! That will be THE day. Amen and Alleluia!!! May God richly Bless you and yours.

A RELUCTANT ICON: LETTERS TO NEIL ARMSTRONG

[Name withheld]

P.S. Pardon the sloppy typing—I hate to type. Be see'in ya!

Besides withholding the name of the letter writer, all names in the letter have been fictionalized. This letter illustrates just how kooky some of the correspondence Neil received was. This man wrote Neil several letters, none of which he or his secretary answered.

"THE AURA YOU EXPERIENCED"

January 19, 1999

Dear Neil:

I enclose a letter from a guide who was assigned to me on a trip to Israel a couple of years ago. While on the ruins of the Temple Mount stairway in Jerusalem, Ori reflected on a sentence you uttered analogizing the similarity of the aura you experienced stepping there and stepping from the Eagle onto the moon. Do you recall the event during a trip to the Holy City?

I promised Ori that I would ask you and I will respond accordingly.

Hope to see you at the spring meeting.

Regards,

David C. Hurley
Chief Executive Officer
Flight Services Group, Inc.
Sikorsky Memorial Airport
Stratford, CT 06497

I have looked in vain in Purdue's Neil A. Armstrong papers collection for Armstrong's written reply to Mr. Hurley's inquiry. I became aware of this story upon reading Thomas Friedman's 1990 book, From Beirut to Jerusalem, *which relates the account as told to him (as with Mr. Hurley) by his guide at the Temple Mount Stairway. In my interviews with Armstrong for* First Man, *I asked him about the story, as I was very interested in documenting and explaining clearly and correctly Neil's religious views. He told me that the story was untrue—a fabrication like so many other stories that had been fabricated about him over the years. Nonetheless, the story about what he allegedly said at the Temple Mount Stairway continues to spread. Currently, a Google search of "Neil Armstrong" and "Temple Mount Stairway" results in 1,140,000 hits.*

2
ANGER, DISAPPOINTMENT, AND DISILLUSIONMENT

Among the hundreds of thousands of cards and letters that Neil Armstrong received over the five decades following Apollo 11, a small fraction of that mail came from people who were disappointed with—or even angry at—his failing to give them what they had asked for in their previous correspondence to him, his apparently behaving in ways they thought were not befitting of him, and any number of other perceived slights the First Man seems to have committed, or shortcomings he exhibited, in his role as public figure and American hero. Some—but not all—of the angriest letters came from conspiracy theorists (the subject of chapter 3) and others who did not believe the Moon landings had truly occurred, who believed that everything about them had somehow been faked (likely by the U.S. government), and that Neil's historic first step—and everything he said or did in his life after July 20, 1969—was a sham, and that Neil knew it. Although the mail Neil received from such individuals amounted to less than 1 percent of the total he received, it always disturbed him to deal with such unbelievers, many of whom were quite belligerent. Most of those he tried to completely ignore; others, who were less malevolent, Neil offered a thoughtful, sometimes even a resounding, reply. But the correspondence that bothered Neil the most came from those who expressed disappointment and disillusionment in him personally; those were the very hardest for him to take. Such letters were far more upsetting, because they were letters written to him not by irrational, misinformed, misguided, or hateful people but rather from everyday normal folks who expressed sincere

ANGER, DISAPPOINTMENT, AND DISILLUSIONMENT

hurt and disenchantment because they felt their hero had in some way let them down.

"PICTURED HERE WITH YOU IS AN UNADULTERATED NAZI"

August 9, 1970

Neil Armstrong:

This ugly woman pictured here with you is an unadulterated NAZI and should have hung at Nuremberg with the others. She would rather pal around with Hitler than with you. It makes me sick that my tax money goes for this this. Don't you care what company you keep?

[Name withheld]
Wilmette, Illinois

Attached to the letter was a photo published in the Chicago Tribune on Sunday, August 9, 1970, captioned "Armstrong in Germany: Astronaut Neil Armstrong, the first man to set foot on the moon, is greeted in Frankfurt, West Germany, by Hanna Reitsch, famous German aviatrix. Armstrong is en route to a glider meet." [Name withheld] has struck the words "famous German aviatrix" and written instead "NAZI WENCH."

Hanna Reitsch (1912–1979) was indeed Germany's most famous female aviator and test pilot. Before and after World War II she set more than forty flight altitude records and women's endurance records in gliders. Reitsch was also the first female helicopter pilot and one of the few pilots to fly the Focke-Achgelis Fa 61, the first fully controllable helicopter. In 1938, during the three weeks of an international automobile exhibition in Berlin, she made daily flights of the Focke-Wulf Fw 61 helicopter inside the large Deutschlandhalle. In September 1938, when Armstrong was eight years old, she flew in the Cleveland National Air Races. She flight tested many of Nazi Germany's latest designs, among them the rocket-propelled Messerschmitt Me 163 Komet in 1942, in which she crashed, injuring herself so badly it required a five-month hospital stay. For her flight testing during the war she was awarded the Iron Cross First Class and the Luftwaffe Pilot/Observer Badge. There is no question that Hanna Reitsch was a Nazi, although very much a politically naïve one. It was not only that she willingly served as an international representative of the fascist regime and one of the most omnipresent stars of Nazi propaganda, but she was an ardent follower of Adolf Hitler. When asked by American military intelligence officers about her capture outside Hitler's bunker in Berlin in April 1945, Reitsch answered, "It was the blackest day when we could not die at our Führer's side." She continued, "We should all kneel down in reverence and prayer before the altar of the Fatherland." She was imprisoned by the U.S. Army for eighteen months before being released. After the war, she went back to flight testing, living much of the time in India and Ghana, where she served as an aviation technical adviser and founded gliding schools. Into the 1970s, Reitsch broke many gliding and helicopter records, some in the United States. It is not clear exactly how much Armstrong knew about Reitsch's wartime connections with the Nazis, though he must have been generally aware of it. It is also possible that he did not know that he was meeting Reitsch during his trip to the Frankfurt air meet in August 1970.

"OBVIOUS THAT YOU HAVE NO INTENTION IN PARTICIPATING IN MY PROJECT"

June 12, 1971

Dear Dr. Armstrong (former astronaut)

I, [name withheld] of Bridgend, Wales, U.K., sent you a letter accompanied with a few of my compositions, one of which I dedicated to you ("American Jamboree").

It now seems obvious that you have no intention in participating in my project or even showing any basic appreciation. I waited, literally in suspense, for a reply from you, but there was no sign.

Therefore, I am forced to "scrubb" you off my list. This, all may seem quite trivial to you and also may not mean anything to you, but to me it means a great deal. I just cannot equate the *first*!! man on the Moon to not replying to a young boy in Wales (U.K.). Little things like this reveal the *true* character of a person. My next astronaut is Buzz Aldrin.

Yours sincerely,

[Name withheld]

P.S. If I ever become an accomplished top astronaut, I will never hesitate to reply to young lads!

P.S.S. A great deal of work went in to my first letter, and my father who enjoyed helping me to prepare the music. I sent the letter and the music by registered post. Just in case that you *haven't* received it, I enclose the registration slip as proof.

"BUT THE AUTOGRAPHS WERE PRINTED!"

December 30, 1971

Dear Mr. Neil Armstrong,

When you were in Oslo, you and your two colleges [colleagues] were so kind to answer my request for your autographs, and so far so good. Thanks anyway—but the autographs were PRINTED! and for a real collector that has no value. Imagine to be a collector of autographs, and then only have a *printed* autograph from the First Man on the Moon! Now I write to you (I should like to write to all three of you) and ask you to be so kind as to present me with your genuine *hand*-signed autograph, preferably beneath a small letter (if your time allows that), maybe *also* a *hand*-signed picture. I should be so happy and grateful for an answer. I am an old man of 52 years (though not feeling like that), but I am deeply interested and impressed by your (and the other astronauts') work, and I should like to have all of you in my collection. BUT YOU FIRST!

Kindest regards,

[Name withheld]
Copenhagen, Denmark

A RELUCTANT ICON: LETTERS TO NEIL ARMSTRONG

"I'VE REALLY BEEN DISTURBED AND ANNOYED"

July 28, 1972

Mrs. Fern Lee Pickens
Assistant to Neil Armstrong
Washington, D.C.

Dear Mrs. Pickens,

I've really been disturbed and annoyed ever since I received your letter of July 5 in reference to refusing a autograph request that I made to Mr. Armstrong.

The reason for this refusal was that he had granted this previously (about five times between October 1970 and December 1971).

Coincidentally, Mrs. Pickens, I have a strong admiration for our Space Program and have been a strong supporter and advocater of continuing this program.

Correspondence from General [Jim] McDivitt advised that to help in this direction, I write letters to my Congressman and promote the program whatever method might be available.

Coincidentally, Mrs. Pickens, I am a serious collector of Space Stamps and Philatelic covers and it dates back to way before Apollo 11. I am not a dealer and I have not and will not commercialise on my hobby. To date I have not and will never sell any of the items I possess. I am a serious admirer, supporter and collector of the U.S. Space Program.

Coincidentally, Mrs. Pickens, I speak before many youth groups in the course of a year—Boy Scouts, Church Groups, P.A.L. (Police Athletic League), Girls Organizations and any youth organization who would like to have my services on this topic. Of course, there is no fee for this. After these talks I ask the questions: Who was the first man to fly the Atlantic? And who was the first man to set foot on the Moon? Believe it or not, many kids do not know the answer to either of those questions. Of course part of my programs, I give away Space Coins and of course the autographed covers stimulate the youth and might motivate this young person to get interested in the Space Program. And those thousands of requests you are complaining about possibly could have been motivated by my talk. I do tell them how easy it is to get started

ANGER, DISAPPOINTMENT, AND DISILLUSIONMENT

and I am sure some of my young people sent in a request for one thing or another. (I didn't know this was a burden, I actually thought I was helping by stimulating a interest in the youth.)

Coincidentally, Mrs. Pickens, I pay for all the covers and all the mailing. It's not a lot of money, but at the end of the year it does add up and there is no commercialization involved. I will admit I get tremendous satisfaction out of it and NASA has send me photos to distribute when I notified in advance of a planned talk—and of course I thank them for it.

Coincidentally, Mrs. Pickens, if there has been any commercialization that needs cleaning up, I suggest you start at home. This latest hit with Apollo 15 and Apollo 14 and so on has infuriated me to no end.

I who plans an album of historical significance, who trys to spread the word so that the Space Program continues cause I strongly believe it, who has a sincere educators viewpoint, who is a strong supporter of this program, gets put down for a simple inscription that might excite a collector or start some high interest in the Space Program, while thousands are working out schemes for hundreds of thousands of Dollars that the taxpayer actually pays for.

I wonder if you sent them a polite letter and threw in a couple books to keep them happy.

I'm sorry. I have strong feelings in this matter as you can see by this long letter which probably won't even get read.

O.K. I said it, and you must admit I got a point, sort of gives me mixed emptions as to what to do next.

Sincerely,

Lou Newman
Liberty, New York

This letter is from the same Louis Newman of Liberty, New York, whose series of letters to Armstrong from 1970 and 1971 requesting various autographs appear in Dear Neil Armstrong: Letters to the First Man from All Mankind.

A RELUCTANT ICON: LETTERS TO NEIL ARMSTRONG

"I WOULD LIKE TO THINK THAT I HAD THE COURTESY OF A REPLY"

January 22, 1973

NASA Headquarters Information Center
To Neil A. Armstrong

Sir:

On November 10, 1972, or more than two months ago, I mailed to you, a return receipt requested, a copy of "First Man on the Moon," by Armstrong, Collins and Aldrin. As this has not been returned to me, or any word received as to its disposition (I requested in the letter enclosed with the proper amount of postage for its return to me at the address below, an autograph of one Neil Armstrong, failing that one or the other of the co-authors would do.)

I understand from a recent TV story, that Mr. Armstrong does not care to have anymore to do with the public; but in view of the fact, that my husband was one of the plastic mold designers for the Deutsch Company connectors used in their helmets and on their suits, I would like to think that I had the courtesy of a reply. The book was not particularly valuable, I think though it was a first edition, which I could not probably obtain again. However, I wonder how long it does take for a book to be returned if no signature is available? Perhaps it and the request have been gathering dust in your mailroom or in your files? My letter which was enclosed in the cardboard wrapper which was then wrapped in brown paper, had more than sufficient postage stuck to both. I asked also that the wrapper and envelope be returned. Perhaps the book has been forwarded to another address? Perhaps the book is still with you waiting for a visit from one or the other of the Astronauts whose autographs I requested?

Would you be so kind as to advise me as to the proper procedure or another address to which I could write for information concerning my book mentioned above?

I expect to inquire of the NASA Manned Spacecraft Center at Code CD, Houston, Texas, 77058 if the book was sent on to them.

Thank you for a prompt reply. I am

Yours truly,

ANGER, DISAPPOINTMENT, AND DISILLUSIONMENT

Mrs. Eugene Robert Bankey ("Jenny" Bankey)
Banning, California

"DISAPPOINTED WHEN I DID NOT RECEIVE THE SIGNED QUOTATION"

September 24, 1974

Dear Professor Armstrong:

For the autographed picture I would like to thank you most sincerely. I assure you that I shall treasure this memento always and share it with many generations of students to come.

Yours sincerely,

Hans-Georg Gilde
Professor of Chemistry
Marietta College
Marietta, Ohio

P.S., I was very disappointed when I did not receive the signed quotation, for it was my intention to frame it and display it—along with quotes from other scientists—in our chemistry building, in the hope to inspire some of our students.

"UNABLE TO HONOR ANY REQUESTS FOR PERSONAL MEMENTOS"

June 12, 1975

John R. Virus
Flushing, New York

Dear Mr. Virus:

Thank you for your letter offering Professor Armstrong the opportunity to participate in your fund raising project by contributing several autographed photos.

A RELUCTANT ICON: LETTERS TO NEIL ARMSTRONG

Professor Armstrong does not knowingly provide his signatures for sale or auction. He is also unable to honor any requests for personal mementos. He receives so many similar requests that in fairness to all concerned it is necessary to decline them all.

We do appreciate your interest and hope that your project will be a success.

Sincerely,

(Miss) Luanna J. Fisher
Secretary to Professor Neil A. Armstrong

A few weeks later Miss Fisher received a copy of her letter back from Mr. Virus with the following note from him typed at the bottom:

June 24, 1975

Dear Miss Fisher,

First of all, I would like you to know that I had nothing whatsoever to do with the WNET Channel 13 FUND RAISING. I was one of the viewers who had noticed personalized items being auctioned off and thought that NEIL ARMSTRONG'S autographed photo would bring in some money for *educational* TV, and it is the reason I wrote in, Merely a "Good Samaritan."

I am rather surprised at his attitude. First of all, he is earning his living at an EDUCATIONAL institution. Secondly, an autographed photo costs *him nothing*. Thirdly, he is a celebrity not through his own merits but through a *public project* which all Americans paid for, and having this status has a *responsibility* to the Americans. He is profiting by this status from recent articles I have read, and I know that he has *never* refused anything given to him. He certainly sounds like a very selfish individual when he cannot "give away" his SIGNATURE. I would hate to depend on him in a time of crisis! His trip through space has taught him very little!

ANGER, DISAPPOINTMENT, AND DISILLUSIONMENT

"YOU HAVE AN OPPORTUNITY TO HAVE A GREATER IMPACT ON SOCIETY"

April 10, 1996

Dear Mr. Armstrong,

I am writing because I am concerned. I'm not sure if you realize that you have an opportunity to have a greater impact on society NOW than you did 30 years ago when you where an intrical part of the NASA space program, and the heritage of our country.

My son Alex (who is 8-1/2 years old) read all about you, and saw pictures and video clips of your accomplishments when he was surfing the World Wide Web back in December. He wrote you a fan letter asking you for your autograph, and received a form letter back saying that after 35 years of doing this kind of thing you where no longer doing it anymore. He was crushed, and moped around for a couple weeks. Now he is back to wanting to be a pro football player and worshipping Emmitt Smith. To look at him you would say everything is OK, but I beg to differ. As a parent I would have much rather had him idolizing you than idolizing Emmitt Smith.

Your past accomplishments have opened up an avenue for you to have an impact on the millions of kids who will take this country into the 21st century. I don't claim to understand why you would no longer want to go down this avenue that God and your hard worked have openned up for you, but as a parent who thinks that today's cry baby, spoiled brat athletes should not be role models, seeing my kid get so EXCITED about something other than a sports personality (he really flipped out over you, the other astronauts, and the various other NASA related topics he read about on the Web) gave me a glimmer of hope for the future of our planet.

I don't want to end this letter sounding like a whiner, so I guess all I can do is ask you to please take advantage of the opportunities you still have ahead of you to "change the face of our country" by touching the minds of our children. Thanks for a lifetime of great accomplishments and good luck with whatever the future holds for you.

Sincerely,

Thomas R. Wolf, Jr.
Portage, Michigan

A RELUCTANT ICON: LETTERS TO NEIL ARMSTRONG

"PACK OF LIES AND DECEIT"

August 1, 2000

Dear Mr. Armstrong:

The least I could do was send a card for your 70th birthday, however over 30 years on from the pathetic TV broadcast when you fooled everyone by claiming to have walked upon the Moon, I would like to point out that you, and the other astronauts, are making yourselfs a worldwide laughing stock, thanks to the Internet.

Perhaps you are totally unaware of all the evidence circulating the globe via the Internet. Everyone now knows the whole saga was faked, and the evidence is there for all to see. We know the pictures have pasted backgrounds, who composed the pictures, and how the lunar landing and Moon walks were simulated at Langley Research Centre, in addition to why NASA faked Apollo.

Maybe you are one of those pensioners who do not surf the Internet, because you know precious little about how it works. May I suggest you visit [website withheld] to see for yourself how ridiculous the Moon landing claim looks 30 years on.

As a teacher of young children, I have a duty to tell them history as it truly happened, and not a pack of lies and deceit.

[Name withheld]

Neil received this letter on his 70th birthday, August 5, 2000. He sent the birthday card and letter on to NASA's associate administrator for policy and plans with the following note: "Has NASA ever refuted the allegations or assembled information to be used in rebuttal? I occasionally am asked questions in public forums and feel I don't do as good a job as I might with more complete information." Subsequently, in 2002, NASA commissioned distinguished space writer and veteran UFO debunker James Oberg to write a 30,000-word monograph refuting the notion that the Apollo program was a hoax. After news of the plan for Oberg's book hit the papers, however, NASA quickly reversed course, judging that not even a judicious, well-argued refutation could successfully achieve its intended effect. Interestingly, Jim Oberg, in October 1970, had sent Neil the following letter as a young man soon after finishing NASA-funded graduate work and moving to Kirtland Air Force Base in New Mexico on active duty with the U.S. Air Force.

October 4, 1970

Dear Mr. Armstrong,

I hope you enjoyed your trip to the USSR; I was attempting to follow it through reading of Soviet newspapers I received, but you were, unfortunately, a very minor story, except when placing wreaths and such. I was in Moscow in 1968 and remember many of the places you were at, but being just a student, I had no chance of getting near Stellar Village. While you were there, what kind of tour did you have? Did you meet any of the cosmonauts and engineers who have not yet flown in space? Did you get any advance notice of Soyuz-9, or are you at liberty to tell?

We had all hoped you would get a ride to Baikonur, but it seems that was not to be. Maybe next time.

You have been quoted as indicating you expect to return to flight status after a few years in Washington. Is that correct? The space stations of the end of this decade will certainly need experienced astronauts to command them. Is that what you had in mind?

I don't want to take up your time, when you're getting ready to move and all, so just this: best wishes and success at all your future endeavors.

All the best,

James E. Oberg
Albuquerque, New Mexico

Interestingly, Mr. Oberg, considered now for many years one of the truly informed and thoughtful experts on the Soviet and Russian space programs, has no recollection of writing this letter to Neil, or receiving anything back, which he did, just a few weeks later, on November 4, 1970. Unfortunately the letter to Jim Oberg was not written by Neil, because it would have very informative to read his responses to Oberg's questions; rather, the response came from Neil's assistant at NASA Headquarters, S. B. Weber, and was only the standard form letter thanking him for his letter and stating that "because of his heavy schedule of activities with NASA, he is unable to respond personally to each letter he receives."

A RELUCTANT ICON: LETTERS TO NEIL ARMSTRONG

"THE VIDEO SUGGESTS THAT YOU DO NOT WANT TO LIE ANYMORE"

October 17, 2005

Mr. Armstrong,

Hello, my name is James Whitman, and I am Social Studies teacher at Sidney High School. I am trying to find NASA Astronaut, Neil Armstrong through an exhaustive search. I hope you are THE man, if not, I am sorry for any confusion. My educational career has led me to pursue elements of living history in an effort to generate student interest in our subject matter. This quest has led me to make contact with you in hopes that we can discuss some extraordinarily sensitive issues. I sincerely apologize for the intrusion to your privacy, but my curiosity is driven by a dedication to my profession.

Born in 1969, I was raised in an era of space exploration. My impressionable young mind dreamt of life in outer space, fueled by the science fiction of George Lucas and Isaac Asimov. My father would take me to the Air Force Museum where I gravitated to the Space exhibits. Though I was always fascinated with subjects like mechanical engineering and astronomy, I answered the call to teaching. My colleagues know I use a slightly unorthodox approach to make learning exciting while developing each student's critical thinking skills.

My government studies explore a unit on Conspiracies in Government which focuses on three major events: the Kennedy Assassination, Watergate, and the Iran-Contra Scandal. But over the past few years, students have challenged me with a most unusual conspiracy involving the Moon landings. I had never doubted that man had reached the moon, despite my Granny believing that it was all phony. As I investigated the questions posed by conspiracy "nuts" and after I placed the moon landings on a timeline that coincided with the corruption of the Nixon Administration, I found myself doubting. Such a thought seemed so disrespectful to men like you, our Aerospace Heroes who risked their lives to achieve such feats. Still, I was compelled to dig deeper. I ordered several videos and purchased several books on the subject from both sides of the debate. The FOX network broadcasted a generic program posing a few questions about the possibility that the moon landings were a hoax. The science community reacted with a

ANGER, DISAPPOINTMENT, AND DISILLUSIONMENT

website (www.badastronomy.com) to debunk the conspiracy theorists. The debate rages on with NASA saying very little for fear that it will be twisted by conspiracy theorists. Their silence only raises more speculation. So it is reasonable for one to ask, "What is the truth?" Can any scientist or astronomer criticize the objective, logical analysis of evidence? I am trying to maintain my skepticism and think logically, and that is why I need your help.

One conspiracy video features a cryptic speech you offered to young students touring the White House in 1994. Your words, "Today, We have with us a group of students, among America's best. To you, we say we have only completed a beginning ... we leave you much that is undone. There are great ideas undiscovered, breakthroughs available to those who can remove one of truth's protective layers." Was there any hidden meaning to this message, like the video tries to suggest? And why do you shy away from celebrating events at the Armstrong Museum located in your hometown of Wapakoneta? The video suggests that you do not want to lie anymore, but perhaps, you are avoiding the limelight for your own personal reasons.

Another conspiracy video, *It Was Only a Paper Moon* (by Jim Collier) analyzes NASA footage and challenges the authenticity of the video when launching from the moon, the robotic movement of the LEM with only a few thruster jets ... paralleling it to cheap Hollywood special effects. It does seem to move robotically as it approaches the orbiter, and I wonder how this is possible with very limited gravity. The Van Allen radiation belts represent another obstacle, as well as the logistics of spacesuit technology mentioned in this video. Please bear with me, there is more.

The most confusing evidence for a conspiracy came with one video, *A Funny Thing Happened on the Way to the Moon* (2001 video directed and produced by [name withheld], the same guy who had an "altercation" with Buzz Aldrin) the video concludes with several clips of what they claim are NASA edits from your moon mission. These clips appear to suggest that your crew was in low-earth orbit while faking a video of being halfway to the moon ... by simply placing a cardboard insert into the round portal window (covering half of the window as a way to show part of the earth and the day/night separation) then moving the camera to the rear of the spacecraft. Later the clip shows a diffused worklight and the insert being removed with the earth in low-earth orbit. The clips include an audiotape of you declaring to be 130,000 miles out. The clips have NASA bluescreen edits dated July 18, 19, and 20, 1969. I cannot find anyone who can explain these video clips or from where they came. Are they authentic, and is there any reasonable explanation for them?

My questions have exasperated most of those around me, as I am often persecuted with jokes for merely questioning things. When one of my colleagues, a Physics teacher, agreed to examine this evidence, he quickly became flustered. His refuting evidence is logical, but much of it is based on the video and photographic evidence, all of which could have been faked. He simply cannot accept the idea that such a secret could be kept or that we Americans could be so gullible, but I am not so sure.

In many ways, I feel so ashamed by all of this doubt, knowing that I may be *"vandalizing history"* while disrespecting men who risked their very lives in pursuit of knowledge. Please forgive me, but as a true American patriot, I think I have ample cause for speculation. I have ample cause for speculation when it comes to the "official version." So my skepticism is fueled by factual conspiracies, and I am seeking the truth wherever I can find it. I need to know if I am teaching fact or fiction. These young minds are seeking the truth from me, and I am not sure I know the truth anymore. Please help me get it right. Would you be so kind as to consider sharing with us a letter of reply or better yet, accept our personal invitation to visit with a small group of students?

Again, I apologize for the intrusion. I greatly appreciate your time and consideration.

Sincerely yours,

James Whitman
Sidney High School
Sidney, Ohio
[Email address withheld]

At the bottom of Mr. Whitman's letter were the following two quotes in italics:

"There is no greater evil than for a man to believe something so completely true, when in actuality, that something is completely false." ~ ~ Unknown

"It is one thing to show a man his position is inferior, and another to put him in the position of truth." ~ ~ John Locke

ANGER, DISAPPOINTMENT, AND DISILLUSIONMENT

"I trust that you, as a teacher, are an educated person"

November 10, 2005

Email to James Whitman from N. A. Armstrong
Subject: Conspiracy

Mr. Whitman,

Your letter expressing doubts based on the skeptics and conspiracy theorists mystifies me.

They would have you believe that the United States Government perpetrated a gigantic fraud on its citizenry. That the 400,000 Americans who worked on an *unclassified* program are all complicit in the deception, and none broke ranks and admitted their deceit.

If you believe that, why would you contact me, clearly one of those 400,000 liars?

I trust that you, as a teacher, are an educated person. You will know how to contact knowledgeable people who could not have been party to the scam.

The skeptics claims that the Apollo flights did not go to the moon. You could contact the experts from other countries who tracked the flights on radar (Jodrell Bank in England or even the Russian Academicians).

You should contact the Astronomers at Lick Observatory who bounced their laser beam off the Lunar Ranging Reflector minutes after I installed it. Or, if you don't find them persuasive, you could contact the astronomers at the Pic du Midi observatory in France. They can tell you about all the other astronomers in other countries who are still making measurements from these same mirrors—and you can contact them.

Or you could get on the net and find the researchers in university laboratories around the world who are studying the lunar samples returned on Apollo, some of which have never been found on earth.

But you shouldn't be asking me, because I am clearly suspect and not believable.

Neil Armstrong

A RELUCTANT ICON: LETTERS TO NEIL ARMSTRONG

"I am truly sorry for offending you"

November 10, 2005

Email to N. A. Armstrong from James Whitman
Subject: Conspiracy

I am TRULY sorry for offending you. That was certainly not my intention. I know you must be weary with the endless questions regarding your personal integrity. I don't know you personally. I'm just examining the facts wherever I can find them. By contacting you, I had hoped to establish a perspective of your character for myself based on your response to my questions. Honestly, I didn't think I would ever receive a response, and I am sincerely thankful for your suggestions.

    Please understand that I tried to contact knowledgeable people first. I sent out a lot of email to reputable observatories and space research facilities around the world (MLRS, NEAT, STSC, Uppsala Astronomical Observatory, Cote de Azur Observatory, the European Space Agency, the Danish Space Research Institute, and the Royal Astronomical Observatory of Canada, to name a few), asking for information on the moon landings, inquiring about the Van Allen radiation belts, etc. I received only three responses and two were basically referrals that also met with no response. My only meaningful response came from a professor at Louisville University, who took the time to share his professional opinion that the Van Allen radiation belts would not represent an obstacle for astronauts at high speed. Please understand, it was not for my lack of trying. I will follow-up on your recommendations to contact Jodrell Bank and the Pic du Midi and Lick Observatories.

    I never meant to come off as some conspiracy nut who is hopelessly attached to a distrust of government. As I said in my letter, I don't want to be some "historical vandal." I was just looking at all the evidence, and it generated some doubt. The conspiracy theorists have presented some very interesting questions, suggesting that the Lunar Range Reflector may have been placed there by one of the Ranger, Surveyor, or Lunar Orbiter spacecraft that studied the moon in the years preceding your mission. I thought the idea had some conceivable possibility.

    Your comment about 400,000 co-conspirators is legitimate and gives me valuable support to believe the landings actually occurred. But the conspiracy theorists are also quick to point out the

ANGER, DISAPPOINTMENT, AND DISILLUSIONMENT

compartmentalization of NASA, where only a limited number of people have access to every aspect of the program. My engineering contacts at WPAFB [Wright Patterson Air Force Base] and Glenn Research Center demonstrate that point. Both individuals are adamant that there was never any hoax, but at the same time, neither can provide categorical proof that it was for real. They saw the same grainy images broadcast to the world. And thus it left me with more reasonable doubt.

I tried the same with lunar rock researchers . . . same result . . . more doubt. You stated, ". . . laboratories around the world are studying lunar samples returned on Apollo, some of which have never been found on earth." The word "some" implies that some similarities have been found on earth and conceivably other samples have not been found . . . yet. Conspiracy theorists suggest that some samples were returned from the earlier moon probes, or the theorists throw out nonsense about researchers taking deep-ocean basalt and pulverizing it with microwaves to "fabricate identical lunar samples." It all adds up to more doubt and speculation for me. What do I know about moon rocks . . . absolutely nothing.

I do consider myself an educated, rational person. I continue to pursue that categorical proof, one way or the other. I wish the Hubble Space Telescope could zoom in and show me once and for all, but it cannot. The camera on HST at its best resolution is .03 arcsec, but to view the landing areas on the moon, it would necessitate a resolution of .002 arcsec. It's just one of many other inconclusive results, results that fuel the debate for conspiracy theorists.

So Mr. Armstrong, I was just seeking clarification so that I can be certain I'm teaching my students the real history. That's why I asked the questions. Obviously, you are by no means obligated to answer anything, and I know I've probably wasted enough of your time. But perhaps, you can understand the growing sentiments of doubt raised by others who are seeking the truth, to confirm or deny the government's official version. Granted, there are some real whackos out there . . . I don't want to be considered one of those folks. If I had reasonable explanations and some categorical proof, I could quell this tide of suspicion. At present, you, Mr. Aldrin, and Mr. Collins seems to be the only sources for proving mankind has walked on the moon.

I will follow your recommendations and consult with those experts. I apologize for the intrusion into your life, and I will not bother you again. Thank you for your service to our country. I hope your Veterans' Day is full of blessings.

Sincerely yours,

James Whitman

Neil did not answer this email, and there seems to have been no additional correspondence of any sort between Neil and Mr. Whitman. Interesting, Sidney High School is located only twenty-one miles down Interstate 75 from the Armstrong Air and Space Museum in Wapakoneta, Ohio, where Neil himself went to high school.

One wonders if the imagery from NASA's Lunar Reconnaissance Orbiter, launched in June 2009, persuaded Mr. Whitman that the Apollo Moon landings were real. The LRO probe made a 3-D map of the Moon's surface at 100-meter resolution and 98.2 percent coverage (excluding polar areas in deep shadow). Its imagery included 0.5-meter resolution images of all the Apollo landing sites. The first images from LRO were published in July 2009.

Also, it is curious that Mr. Whitman believed that the Apollo 11 crew—Neil, Buzz, and Mike—were "the only sources for proving mankind has walked on the moon." Five other Apollo crews made landing missions (Apollo 12, 14, 15, 16, and 17) and three others circumnavigated the Moon (Apollo 8, 10, and 13)—the last on the list, of course, was meant to be a landing mission but had its landing aborted after an oxygen tank exploded two days after launch, crippling the service module upon which the command module depended. Mr. Whitman could have thus asked his questions of any of those other crew members, particularly Jim Lovell (Apollo 8 and 13), John Young (Apollo 10 and 16), or Gene Cernan (Apollo 10 and 17), who each made trips to the Moon on two occasions.

Mr. Whitman's correspondence to me dated March 14, 2018, in which he grants permission to include his two letters in this book, closes with the following statement: "Today, the material I've generated to guide my classes represents a careful analysis of evidence as well as an examination of arguments from both sides to help students understand the lure of conspiracy theories. I hope that encouraging students to have the healthy skepticism of an open mind is always balanced by keeping them firmly grounded in facts."

3
QUACKS, CONSPIRACY THEORISTS, AND UFOLOGISTS

It would be nice if it were not necessary to include a chapter in this book on quacks and conspiracy theorists. But the unfortunate truth is, Neil Armstrong over the years received a significant number of letters from charlatans who pretended to be other than they were (Neil labeled them "quacks"), diverse oddballs, and people who must have been clinically demented or mentally deranged (Neil labeled them "crazies"). Such letters are the bane of most famous people, and many such letters can be simply dismissed or ignored. In other instances, the person writing may also have been a stalker and constituted a possible physical danger. In these cases, precautionary measures needed to be taken, typically by alerting law enforcement and other responsible officials of the threat.

Without taking such letters into account, the iconography involving Neil Armstrong, sadly, is not complete. Still, to give too much voice to such letter writers would also be a mistake, perhaps inciting other unsound minds to follow in the footsteps of Neil's quacks and crazies. Thus, only a few examples of such letters are included in this chapter, and in several cases the author name and address are withheld. Still, it is noteworthy that Neil kept all such letters for a reason, and they will forever be part of his legacy as part of the Neil A. Armstrong papers collection within the Purdue University Archives and Special Collections' larger Barron Hilton Flight and Space Exploration Archives.

Also included in this chapter is a sampling of letters Neil received from fans of unidentified flying objects. Although Neil did not regard everyone interested in the possibility of UFOs as quacks or crazies, he did believe most of them were just that and did what he could to pour cold water on most UFO speculation. Typically he would be asked about the UFOs the Apollo 11 crew allegedly encountered. According to the various stories, on their way to the Moon the Apollo 11 astronauts saw something, perhaps several things, they could not identify, ranging from mysterious lights to actual formations of spaceships. One story had a "mass of intelligent energy" tailing the spacecraft all the way from Earth to the Moon. Another had a "bright object resembling a giant snowman," which later "proved" to be two UFOs, racing across Apollo 11's path as it reached lunar orbit. Still another had Mike Collins spotting a UFO as he was photographing the lunar module's ascent from the surface. Allegedly, close analysis of the Apollo 11 film and photographs verified the presence of UFOs, but NASA, the Pentagon, and the entirety of the U.S. government all conspired to cover up the evidence, going so far as to edit out major portions of Apollo 11's onboard and air-to-ground audio recordings. Most incredible of all, a story made the rounds that just as Armstrong started down the ladder to make his historic first step off the LM and onto the Moon, he saw something on the lunar surface that made him scramble back into the LM, only venturing out again several minutes later after Mission Control calmed him down and insisted that he and Aldrin make their Moon walk.

As Neil understood, in broad sociocultural perspective, it is not at all surprising to find that people fascinated with UFOs and the possibility of extraterrestrial life projected their anticipations onto the astronauts and their missions. The sighting of "flying saucers" had reached epidemic proportion in the years following World War II, likely brought on by the specter that came with the appearance of atomic bombs and a need to externalize fear. So crazy had the belief in UFOs become in the late 1940s that the U.S. Air Force began to amass hundreds of case studies and firsthand personal testimonies about UFOs from pilots and common folk alike, in what came to be called the "Blue Book investigations"; in the end, the USAF's conclusions satisfied no one. The UFO craze crested in the late 1950s and early 1960s with the birth of the Space Age, heightened by the launch of the first satellites and space capsules. Any comment about a possible unidentified flying object, especially if made by a pilot, set off another string of reported sightings. Stories circulated that NASA test pilot

QUACKS, CONSPIRACY THEORISTS, AND UFOLOGISTS

Joe Walker, in April 1962, filmed five cylindrical and disk-shaped objects from his X-15 aircraft; that two radar technicians, in April 1964, watched UFOs following an unmanned Gemini capsule; that NASA installed a special instrument on Gemini IV to detect UFOs; that Frank Borman and Jim Lovell on Gemini VII and Pete Conrad and Dick Gordon on Gemini XI had spotted "bogeys." Whatever it was that was originally factual behind the reports quickly got lost amid the illusions, gross exaggerations, and outright fabrications that fed the public's growing appetite for news about UFOs.

It would have been astounding if something as epochal as the first Moon landing had not generated a fresh and intense new round of UFO stories even more hyperbolic and stubbornly persistent than what came before. On the internet today, a Google search for "Apollo 11" in combination with "UFO" results in 1.16 million hits, "astronaut" and "UFO" in 848,000 hits, "pilot" and "UFO" in 1.34 million hits, and "UFO" alone in 635 million hits. Just for "Neil Armstrong" and "UFO," one gets 449,000 hits, for Buzz Aldrin, 378,000, and for Mike Collins, 527,000. Clearly, it is important to those who want to believe in alien intelligence that the First Man on the Moon and his two crewmates spotted a UFO.

Later in this chapter, in a reply to a question about the UFO purportedly experienced by Apollo 11, Neil had his secretary Geneva Barnes explain, very succinctly, based on a statement that he had provided her, what his crew saw and what he believes the UFO to have been.

"A PROPHETIC KEY TO WORLD HISTORY"

November 23, 1969

Dear Sir,

I had the fortune to see and hear you and your friends as you passed through Sydney recently. I was particularly interested in the close parallel between Jules Verne's fiction story a hundred years ago and the actual Moonflight of Apollo 11. This has encouraged me to write you and share some further remarkable data and especially the parallel between you and Columbus.

A RELUCTANT ICON: LETTERS TO NEIL ARMSTRONG

Apollo 11 Name Composition	Apollo = A Columbia = C Eagle = E	Michael = M Edward = E Neil = N	Thus: ACE MEN
Archaeological duration: The event took place 20–21 July 1969 (world)	a) on Sunday the Sun's sign b) in Leo the Sun's sign	a) and Monday the Moon's day b) and Cancer the Moon's sign	
The Act (compare Rev. 10)	c) with the Right Foot (related to the Sun) placed on part of Apollo (the Sun)	c) and the Left Foot (related to the Moon) placed on the Moon	

A prophetic key to world history was revealed to King Nebuchadnezzar of Babylon. According to this key the exploration of the Moon coordinates with Columbus' discovery of America.

Using this key to history I forecast—my lectures during 1962 that an American—the "Columbus of Space"—returns from the Moon in 1968. True enough the parallel is so strong that nearly all comments on the Apollo 11 Moon Flight were in terms of a comparison to Columbus' discovery of the then New World.

For example you were pictured together with Columbus in *Time* magazine July 18, 1969. I myself dropped my chin in utter amazement when I learned that the flight operations director's name is actually *Christopher Columbus* Kraft!

I enclose a sketch of the Image Key to History for your own perusal. Of special interest to you is that I forecast new means of spacecraft propulsion from 1980. Rocketry may be compared to the age of the sailing ships and the new methods will parallel the development of the steam engine. Later on in the 1990s we will probably have learnt how to manipulate fields and the propulsion of spacecraft will then be by electronic means.

Is there any truth in the rumours that you and your fellow astronauts have seen strange spacecraft? If you are free to speak on the matter I would certainly appreciate your answer.

The honest truth is that I and some friends of mine were in communication with people claiming to come from the direction of the Andromeda constellation. Am enclosing a photo-copy of an original "papyrus" communication from them. As you can see there are a number of dots and symbols on it. It tells about the transformation of our solar system through the change of Jupiter into a second sun.

This and much more, but I must end here wishing you and the other pioneers of the space age every success.

Blessings a thousandfold.

[Name withheld]
Concord
New South Wales, Australia

"IT WAS SUCH A STRONG THING WITH ME"

July 14, 1970

Dear Neil,

I do not want you to think this letter is foolish. About a year ago during your blast-off and throughout the space journey and especially during the descent to the moon and subsequently your departure from the moon and link up of LEM and command module we all needed great faith.

I prayed many times for your safety and for your crew members. However there was always a strange coincidence. During each prayer, the name Connie, or Consuelo, came so strongly in my thoughts.

In all the tough and tense situations during those times, can you in total recall, remember if the name passed through your mind and thoughts? It was such a strong thing with me, it makes me wonder.

I'd very much appreciate hearing back from you.

F. St. John Armstrong
Oak Forest, Illinois

A RELUCTANT ICON: LETTERS TO NEIL ARMSTRONG

Reply from S. B. Weber, Neil's assistant

October 26, 1970

Mr. F. St. John Armstrong
Oak Forest, Illinois

Dear Mr. Armstrong:

Please accept our apologies for the delayed response to your letter of July 14, 1970. Mr. Armstrong is always pleased to hear from the many friends who "lived through" every moment of the lunar landing mission with the Apollo 11 crew. Unfortunately, because of his heavy schedule of activities with NASA and the tremendous amount of mail that he receives, Mr. Armstrong is unable to make a timely, personal response to each letter.

Mr. Armstrong is unable to recollect the name Connie or Consuelo passing through his mind during the mission. There could be a lot of explanations for such an occurrence, and we hope that you are able to discover the answer to your puzzle.

Best wishes and, again, thank you for your interest.

Sincerely,

S. B. Weber
Assistant to Mr. Neil A. Armstrong

"THE PLANET MAY BE RUN BY OTHER CREATURES THAN HUMAN"

July 29, 1970

Astronaut Armstrong,

I'm sitting in what appears to be Penobscot county jail awaiting a grand jury indictment for robbery.

If it ever gets to trial, I will need your help. The air may be very clean, but it strikes me that the waters of human commerce have indeed been muddied in this place.

So, I am not too anxious to follow the usual procedure and claim insanity for having observed an unusual flight. And it seems, in spite of threats and intimidation by some (terrestrials and/or aliens), neither do some others.

I believe I spotted the crunch down of an important alien flight the night B4 my arrest. The next morning I was hurried away by the people who supposedly owned the land. (I was taking some astronomical observations) told to leave by noon!

My gas tank was MT so I thought I'd hurry matters and alert someone by requisitioning gas in name of the U.S. Big mistake! She (frightened human or cagey alien)? called police and reported "robbery."

So, in the interests of space law, inter-space accord and other things, I need some scientific, commercial, and perhaps military support for the trial—if there is to be one.

I get the feeling if extra-terrestrial interests previal, I'll sit a long time in some nut ward. If not, and the trial comes to pass, I risk imprisonment.

Anyway, I think the issues are too important to let my fate be determined either by "head in the sand" humans or self serving aliens.

If, just if, we have any alien artifices or any kind of proof that such beings exist and have visited us, I think it's time to perhaps de-classify to some degree, and free ourselves from too much secrecy. Of course, I have a personal stake, but if the answer is "no" because of the possibility of "witch-hunting" for aliens, I can understand.

My point is, if the humanity of the planet is not on solid ground about this, the planet may be run by other creatures than human. Maybe there are other choices.

Of course, if the humans here are too frightened by the propositions of this letter or the weapons of the aliens, or the aliens so foolishly bent on "taking over" this letter will never reach you.

God grant that it does. I'd venture to say we all could learn from each other.

Well, perhaps you can't help but some other crew may have even more experience to offer. My lawyer seems stymied, the police hostile, and I'm discouraged.

Sincerely,

[Name withheld]
Penobscot County Jail
Bangor, Maine

A RELUCTANT ICON: LETTERS TO NEIL ARMSTRONG

"THERE APPEARED ANOTHER SUCH HOLE"

January 14, 1971

Dear Mr. Armstrong,

My name is Cathie Newman and I live among the mountains of the Eastern Districts of Rhodesia, which is in the southern portion of the African continent.

On Tuesday Jan. 12th at approximately 3:30 P.M. Rhodesian time the sun was within 2½ to 3 hours of setting. I happened to look up at the sky in the sun's direction. Behold me there was a most beautifully coloured deep hole (the only way I can describe it). It did not strike me as being a cloud, so I called my servant to look as well. There appeared another such hole with the most beautiful colourings exuding. Also what looked like a big ocean, green and blue colouring, like our sea reaters came into view. To the right of this there was black space. At 4 P.M. this began to come into focus, the colouring was fantastic. On this portion was one very big hole with two smaller ones with such colourings as I cannot even vaguely describe. It was so vivid and seemed so close by. Below the big hole there were deep ridges with more ridges further right and left but lower down. There appeared very very high ground with deep deep ravines all in magnificent colouring. I cannot describe all we saw. Further down and to the left there was a dark gray mass with what appeared to be huge rocks.

My African servant and myself were spellbound. We were speechless, and literally stuck to the ground. At 5 P.M. as the sun got lower in the sky the sight just faded leaving only the usual sky we see every day.

Mr. Armstrong, were we looking at the surface of the moon? If so, how did it come about? Was it something unique that could possibly go down in history?

My local paper has taken it up in the hope that somebody else may have seen it.

If, on the other hand, my servant and I were the only ones to see it, perhaps we have seen what scripture has taught—one of the signs of the destruction of our planet. I sincerely hope not.

I must here state I am not at all a religious person. I also do not drink or take drugs. I am a healthy, perfectly normal person.

Again, Mr. Armstrong, I cannot express in words the magnificence of the whole thing and will be very anxiously awaiting a reply from you.

May I take this opportunity of wishing yourself, your wife and family everything you wish yourselves during 1971.

With kindest regards then. I remain

Sincerely Yours,

Mrs. Cathie Newman
Umtali
Republic of Rhodesia

P.S. I omitted to mention during the time clouds came and went completely obscuring our view at times but the vision still remained when the clouds had past.

"WHAT DO YOU MEAN ACROSS THE 'FLYING SAUCERS'?"

January 24, 1971

Dear Mr. Armstrong,

First I want congratulation to your great result during the flight of Apollo 11 and landing on the moon. But I want ask you, what do you mean across the "Flying Saucers"? Before one months I did ask Mr. James McDivitt, because he had seen during his flight with Gemini IV, an object in space which he was unable to identify. The ground stations were also unable to verify the object which he saw. Mr. McDivitt did in his 20 years of flying he had not seen an object which he is sure it came from some other world. How do you think about? Did you also seen such an object?

In March Mr. Lovell, the Captain of Apollo 13, came here to Tyrol at ski-hollydays. Mr. [Ed] White, how sorry is dead, was also here in Tyrol. In June 1970 I did personally meet Mr. John Glenn. He was in Innsbruck, the capitol of the wonderful ski-ing land Tyrol.

Sincerely,

George Funkhauser
Bosdornau, Austria

A RELUCTANT ICON: LETTERS TO NEIL ARMSTRONG

Reply from Geneva Barnes, Neil's secretary

September 13, 1971

Mr. George Funkhauser
Bosdornau, Austria

Dear Mr. Funkhauser:

Mr. Armstrong has asked me to thank you for your letter of last January. I apologize for the delayed response but our volume of mail has been such that we have not been able to respond as quickly as we would like.

The Apollo 11 astronauts did observe an object during our traverse to the moon which is assumed to be, but cannot positively be identified as, the Saturn S-IV-B stage which propelled us to the moon. The S-IV-B stage is subsequently impacted on to the lunar surface.

In spite of the fact that we track with our radar network thousands of satellites and other pieces of space debris such as rocket stages and shrouds, no tracking of any unknown object from outer space into the atmosphere has ever been recorded. Hence, it is highly unlikely that any of the reported or unknown objects can be identified as extraterrestrial visitors.

Sincerely,

(Mrs.) Geneva Barnes
Secretary to Mr. Armstrong

In my biography of Armstrong, First Man, *I cover in detail both the alleged sighting of a UFO by the Apollo 11 crew and their report of seeing "flashes" out the window of their spacecraft,[11] the latter of which was not covered in the reply from Geneva Barnes to Mr. Funkhauser. In later years when Neil received what he perceived to be serious inquiries concerning the alleged UFO, he would sometimes reply by telling the questioner to consult the* Apollo 11 Technical Crew Debriefing, *July 31, 1969, available on the internet. In volume 1, section 6 ("Translunar Coast"), of that originally confidential publication, Armstrong, Aldrin, and Collins explained what they saw, but could not explain with certainty, on the way to the Moon.[12] Given the subject of this chapter, it is worth reproducing that part of the debriefing here.*

ALDRIN: The first unusual thing that we saw I guess was 1 day out or something pretty close to the moon. It had a sizeable dimension to it, so we put the monocular on it.

COLLINS: How'd we see this thing? Did we just look out the window and there it was?

ALDRIN: Yes, and we weren't sure but what it might be the S-IVB. We called the ground and were told the S-IVB was 6000 miles away. We had a problem with the high gain about this time, didn't we?

COLLINS: There was something. We felt a bump or maybe I just imagined it.

ARMSTRONG [to the debriefing team]: He [Collins] was wondering whether the MESA [Modular Equipment Stowage Assembly, on the lunar module] had come off.

COLLINS: I don't guess we felt anything.

ALDRIN: Of course, we were seeing all sorts of little objects going by at the various dumps and then we happened to see this one brighter object going by. We couldn't think of anything else it could be other than the S-IVB. We looked at it through the monocular and it seemed to have a bit of an L shape to it.

ARMSTRONG: Like an open suitcase.

ALDRIN: We were in PTC [passive thermal control, also known as the "barbecue mode," in which the lunar module/command and service module stack was oriented with its long axis perpendicular to the sun so that, when the spacecraft was put into a slow rotation around that axis, heating and cooling would be relatively uniform] at the time so each one of us had a chance to take a look at this and it certainly seemed to be within our vicinity and of a very sizeable dimension.

ARMSTRONG: We should say that it was right at the limit of the resolution of the eye. It was very difficult to tell just what shape it was. And there was no way to tell the size without knowing the range or the range without knowing the size.

ALDRIN: So, then I got down in the LEB [lower equipment bay] and started looking for it in the optics. We were grossly mislead [sic] because with the sextant off focus what we saw appeared to be a cylinder.

ARMSTRONG: Or really two rings.

ALDRIN: Yes.

ARMSTRONG: Two rings. Two connected rings.

COLLINS: No, it looked like a hollow cylinder to me. It didn't look like two connected rings. You could see this thing tumbling and, when it came around end-on, you could look right down in its guts. It was a hollow cylinder. But then you could change the focus on the sextant and it would be replaced by this open-book shape. It was really weird.

ALDRIN: I guess there's not too much more to say about it other than it wasn't a cylinder.

COLLINS: It was during the period when we thought it was a cylinder that we inquired about the S-IVB and we'd almost convinced ourselves that's what it had to be. But we don't have any more conclusions than that really. The fact that we didn't see it much past this one time period—we really don't have a conclusion as to what it might have been, how big it was, or how far away it was. It was something that wasn't part of the urine dump, we're pretty sure of that. Skipping ahead a bit, when we jettisoned the LM, you know, we fired an explosive charge and got rid of the docking rings and the LM went boom. Pieces came off the LM. It could have been some Mylar or something that had somehow come loose from the LM.

ALDRIN: We thought it could have been a panel, but it didn't appear to have that shape at all.

COLLINS: That's right, and for some reason, we thought it might have been a part of the high gain antenna. It might have been about the time we had high gain antenna problems. In the back of my mind, I have some reason to suspect that its origin was from the spacecraft.

ALDRIN: The other observation that I made accumulated gradually. I don't know whether I saw it the first night, but I'm sure I saw it the second night. I was trying to go to sleep with all the lights out. I observed what I thought were little flashes inside the cabin, spaced a couple of minutes apart and I didn't think too much about it other than just note in my mind that they continued to be there. I couldn't explain why my eye would see these flashes. During transearth coast, we had more time and I devoted more opportunity to investigating what this could have been. It was at that point that I was able to observe on two different occasions that, instead of observing just one flash, I could see double flashes, at points separated by maybe a foot. At other times, I could see a line with no direction of motion and the only thing that comes to my mind is that this is some sort of penetration. At least that's my guess, without much to support it; some penetration of some object into the spacecraft that causes an emission as it enters the cabin itself. Sometimes it was one flash on entering. Possibly departing from an entirely different part of the cabin, outside the field of view. The double flashes appeared to have an entry and then impact on something such as the struts. For a while, I thought it might have been some static electricity because I was also able, in moving my hand up and down the sleep restraint, to generate very small sparks of static electricity. But there was a definite difference between the two as I observed it more and more. I tried to correlate this with the direction of the sun. When you put the window shades up there is still a small amount of leakage. You can generally tell within 20 or 30 degrees the direction of the sun. It seemed as though they were coming from that general direction; however, I really couldn't say if there was near enough evidence to support that these things were observable on the side of the spacecraft where the sun was. A little bit of evidence seemed to support this. I asked the others if they had seen any of these and, until about the last day, they hadn't.

ARMSTRONG: Buzz, I'd seen some light, but I just always attributed this to sunlight, because the window covers leak a little bit of light no matter how tightly secured. The only time I observed it was the last night when we really looked for it. I spent probably an hour carefully watching the inside of the spacecraft and I probably made 50 significant observations in this period.

ALDRIN: Sometimes a minute or two would go by and then you'd see the two within the space of 10 seconds. On an average, I'd say just as a guess it was maybe something like one a minute. Certainly more than enough to convince you that it wasn't an optical illusion. It did give you a rather funny feeling to contemplate that something was zapping through the cabin. There wasn't anything you could do about it.

ARMSTRONG: It could be something like Buzz suggested. Mainly a neutron or some kind of an atomic particle that would be in the visible spectrum.

"THESE STRANGE OBJECTS WERE RUSHING OVER MEXICO"

January 25, 1971

Dear Mr. Armstrong:

I deeply appreciate your kind attention in sending me your autographed picture. We will keep it as an honour to our family.

In our home we have a very special room, quite interesting. On boards we have pasted newspaper and magazine scraps with material since the first rockets, first men in space, all the way up to now, the first airplanes, automobiles, when Carles Lindberg came to Mexico "Spirit of St. Louis." I was lucky enough to see him in the American School where I studied; first railroads, submarines, ships, etc.

Another album, our "Blue Book," contains everything we have found relative to UFO's, in Spanish "Ovnis." One afternoon in Sept. 1965, these strange objects were rushing over Mexico, D. F. [District Federal]. Many people saw them. My husband, children and some friends, climbed on the roof to watch them like brilliant white stars speeding through the sky in all directions. We counted 8; it was dark now, about 7:30 when suddenly over our own home some strange object came flying softly without noise. It was oval silvery gray, like a balloon tire, square spaces—not exactly windows—showing red light from the interior. This UFO passed by and a few yards away it lighted a propulsion motor or something similar, which expelled bright light white and red glows. I made an oil painting of what I saw. I show this album and painting to few people cause they might think I am out of my mind or it's my imagination, but at home we all know we certainly saw it. We have seen them more times and every time the electric lights go out. I painted Mr. Ed

White walking in space. Now I would like to watch the Apollo when it starts from earth in Cape Kennedy and paint it too. My family and I some time will come over to Houston and maybe visit NASA, so that I can paint this wonderful scientific place.

We never miss the flights in T.V. and await with interest the Apollo 14, January 21, and pray to God as always that Mr. Shepard, Mr. Mitchel and Mr. Roosa, return safely to earth.

I hope you like the song I wrote for you and have enclosed hereto "Ballad in Space." If you play the piano, try it, it is easy to play.

Forgive me boring you with such detailed letter. Please consider that our home is at your disposal anytime you and your nice family wish to come to Mexico. With best wishes

Sincerely Yours,

Bertha Carlota Valdés de Sola
Amores, Mexico, D.F.

"DEAR ILLUSTRIOUS BROTHER ARMSTRONG"

March 10, 1971

Mr. Neil Armstrong, 33rd Degree
NASA Headquarters
Washington, D.C. 20546

Dear Illustrious Brother Armstrong:

Sometime ago Brother Charles D. (Chuck) Robinson of Sperry Piedmont Corporation in Charlottesville, Virginia, spoke with your secretary by telephone—during which conversation she told him that you would autograph Scottish Rite Patents whenever we would send them to you. All of the brothers whose patents are enclosed, together with their children and grandchildren, will appreciate this gracious and kind favor.

There are eighteen patents enclosed.

Sincerely and fraternally yours,

Linden Shroyer
Crozet, Virginia

P.S. I am enclosing stamps in the amount of $2.18 which I hope will cover return postage.

Reply from Geneva Barnes

April 6, 1971

Dear Mr. Shroyer:

Mr. Armstrong has asked me to return your package of Scottish Rite Patents. He did not autograph the Patents as he is not a member of the Masonic organization. However, Colonel Aldrin is a member and if you would like to send the Patents for his signature, you may reach him at the following address:

Colonel Edwin E. Aldrin, Jr.
Astronaut Office
Manned Spacecraft Center—NASA
Houston, Texas 77058

Sincerely,

Geneva Barnes
Secretary to Mr. Armstrong

Among the conspiracy theories associated with Apollo 11 is that the Moon landing was a conspiracy of Freemasons. The "evidence": that Buzz Aldrin, who was a Mason, carried a Masonic flag with him in his Personal Property Kit (PPK), which he presented upon his return to the lodge's Sovereign Grand Commander of the Supreme Council of the World, which is true, and that Neil's father was a thirty-third-degree Mason. Neil's father, Stephen, was a Mason but Neil did not know what rank his father achieved.

"THE POSSIBILITY OF THEM COMING FROM OTHER PLANETS"

April 23, 1971

Dear Mr. Armstrong,

QUACKS, CONSPIRACY THEORISTS, AND UFOLOGISTS

I wish to congratulate you for your historic landing on the moon two years ago, and I am very much impressed by our nation's space efforts.

I am going to college this September to study aerospace engineering and hope to take part in the design of spacecraft as our country can explore the other planets.

I am also interested in unidentified flying objects and read an article in Saga Magazine in regards to your Apollo 11 crew seeing one in space. Would like to know if you take an interest in the subject and the possibility of them coming from other planets.

I want to wish you the best of luck and am looking forward to hearing from you.

Yours Truly,

Richard Worley
New York, New York

Armstrong answered this young man's letter personally but the substance of his letter was virtually the same as that which Geneva Barnes sent as a reply to George Funkhauser's January 24, 1971, letter earlier in this chapter.

"WHETHER THAT PRESENTS ANY DANGER TO AUTOMOBILES"

July 4, 1971

Dear Sir:

Two years ago, on that great day, I wished you, for your mission, the same good luck I once had, and which I kept a secret. On a winter's night, I watched a heavy shower of stars which was falling for a long period of time. Dear Captain Armstrong, I would like to know whether that presents any danger to automobiles, and how such a shower of stars comes to pass. I confess that I am content when the sky is overcast, and that I am happy when the weather is beautiful.

May I thank you in advance for your answer; however, I would appreciate receiving it in French.

Dear Astronaut, please accept my regards.

A friend.
Sincerely,

Mrs. Claude Voisaud
Pont Andorra
Normandy
France

"WE CAN EVEN SEND HIM BABY PETS"

October 23, 1971

Neal Armstrong:

I am writing to you to ask you and our country for the opportunity to go to the moon. The aspect and request I am most concerned about is the fact that life again, as we know it, must transfer itself for the next consecrated world. To do this I make to you a request; that a son be raised on the moon, to return when he is 17 or 18. No more need be said as to the year of his return and his size. He would have an of course understanding that would be big enough to understand our direction without the aid of impending removal of popularity. Perhaps we can even send him baby pets so there remains may tell a future humanity by there presentence prudence in tampering with life. Our fine statue in Egypt is past and future.

Our grapes riper than ever since having been dubbed in Phoenix the spring of this year, to me is ready to be left, along with a certain number of mankind.

Make me first or last, but please include me if your role in our government is decided to administrate. I am competent in engineering such a dwelling in an apprentice associateship, or by myself.

In any case please include me on your list, and thank you very much for reading this letter.

Sincerely,

[Name withheld]
Phoenix, Arizona

A.e. Saw Glenn and White's companion from the ground, and as far as

I am concerned, you are the third wize man, G.A.W! Would like to see and talk with you on a per Diem basis, in any case please respond to me.

In cursive handwriting the letter writer added the following at the bottom of his letter: "I am your past Jesus Christ. The child will help me administrate. I can cure almost anything if given the chance. DeVere."

"STILL THE HERO OF IT ALL"

July 24, 1972

To the Former Moonwalker's Secretary

Dear Madame Secretary:

I take the privilege to talk first to you, Professor Armstrong doesn't understand the delicatessen of my writings and says, "Does that oddball send me again these crazy handmade cards and where is the joke." Besides he thinks, I am a third grader who can't spell, he doesn't know I never learned the language at all and my English when I speak it sounds like a diver lands flat on his belly with a big smack on the water. He hates me because I write so enormous stupid, NASA said they didn't have room for me in their operation. They didn't to say, "they are afraid to let me fly, I properly fell off the moon."

I tell you, your boss for me is still the hero of it all, he really made it happen. What are beautiful earth longed and dreamed of ever since it's birth. It's only sad the really great man of our time have no time for the little everyday people to tell them a few nice words. I am one of them to look-up to a man like Armstrong to have made history—history not equal in its greatest power on this earth. Somewhere, somehow it is connected with God, and closer to God than any other person. I think Mr. Armstrong doesn't understand either that he still is adored by people, so from me, he thinks it's crazy. Well every fan has its different way of expression. I am always unlucky in expressing myself. I properly can be glad not to make famous persons throw fits that I take there time.

Now, dear Secretary, would you be good enough to tell Astronaut Armstrong (for me is always *the* astronaut who kept that earth of ours in the greatest excitement with his moon, our moon) that I hope they captured his beautiful heartwarming smile in the museum. The famous

Neil Armstrong smile as known and important as his toes sticking in the moon dust. I just love that good open smile. I have all sorts of pictures from Astronauts but none and the most important from Neil. I have his birth certificate but I don't like him as a baby around me. I am an old odd spinster without a car. I properly never get even to the Museum in Wapakoneta. What good do me invitations when I never have transportation.

Also would you kindly present Mr. Armstrong's birth certificate to him. I find one bad mistake on it. Could that mistake be the explanation for that strange newspaper pictures of band? He is such a good looking guy, maybe there was a jitterbug in the camera. That doctor just lived on the street, he didn't live in a city.

Maybe you can dictate your boss a little letter to me and make his appointment book with "call to say to a little people just . . . hello . . ." that's just one small step when he is around. Gosh. I and others worried more than a President or the Great Society. Does he know that NASA neglected Germany!! Astronauts were sent on good will tours everywhere but never to Germany.

Astronaut Armstrong should have made ever so often publicity—always—to keep in contact with the folks.

Thank you dear Secretary. You tell him will you please.

With best wishes for Neil Armstrong and his lovely family.

Respectfully yours,

Inga Rost
Columbus, Ohio

Ms. Rost was wrong about the astronauts not making a goodwill tour to Germany. As part of the "Giant Step" tour in October 1969, the Apollo crew and its entourage did, in fact, make stops in Cologne and Berlin.

"WENT OUT OF MY MIND"

February 16, 1974

Dear Secretary to Professor Armstrong:

Please forgive me for all those silly things I have written to Professor Neil Alden Armstrong. Those silly things are *NOT* true. The reason why I wrote them is because I got carried away and went out of my mind. *I did not* mean to waste ink, stamps and letters. I am very sorry for what I have done. I will not do it again. Please tell Professor Armstrong I got carried away and I am sorry for doing these silly things. Please let Armstrong read the next paper which is personal.

With best wishes,

Algherd Akula
[Location unspecified]

"THE INFINITE POSSIBILITIES OF MAN"

May 27, 1975

Dear Mr Neil Armstrong:

Am in the process of preparing an article (or lecture: An astrological correspondence between Perry, E. Hilary, Thor Heyerdahl, & Neil Armstrong. The main theme being the infinite possibilities of man.
    In order to do this correctly, I would need the correct time of birth (on birth certificate). By a quick calculation it seemed to be approx. 9:30 A.M. IS THIS CORRECT??? A condensed report will be submitted for a reply as the Moon landing by far eclipses any human endeavor so far.

Sincerely,

Warren Hardy
Santa Monica, California

At the bottom of the letter in Neil's cursive handwriting is the comment "No answer." Neil was categorically not a believer in astrology. Interestingly, his secretary did respond.

A RELUCTANT ICON: LETTERS TO NEIL ARMSTRONG

Reply from Luanna J. Fisher

June 4, 1975

Warren Hardy
Santa Monica, California

Dear Mr. Hardy,

Professor Armstrong has asked me to thank you for your letter and tell you that he does not have the exact time of his birth.

Sincerely,

(Miss) Luanna J. Fisher
Secretary to Neil A. Armstrong

Neil was born at 12:31:30 on August 5, 1930. His mother, Viola, certainly knew the exact time of her first child's birth, and Neil, if he had wanted, could have asked her. He must have known from his mother the general time that he was born.

"Closest approximation known"

June 30, 1975

Dear Miss Luanna J. Fisher,

Thank you for your reply of June 4, 1975. I realize that very few persons have the exact birth time. What was desired was the CLOSEST APPROXIMATION KNOWN of Prof. Neil Armstrong's time of birth.

It is unfortunate that biographical information is missing for many of the important figures in history; a loss for future generations.

Sincerely,

Warren E. Hardy
Santa Monica, California

QUACKS, CONSPIRACY THEORISTS, AND UFOLOGISTS

"DONATE US THE $40,000.00 WE WERE CHEATED OUT OF OUR SAVINGS"

October 10, 1975

Dear Professor Armstrong,

I studied in Germany to become a Caritas nurse and after passing my examination I came to New York.

Here my sister and I have been badly cheated through an effective trick. The trickster pretended that he needed to give money to help someone of his friends for 4 months. Using this deception he borrowed $68,000.00 from us which he deposited in his name. He then squandered away 2/3 of it, then committed suicide.

Both my sister and I have worked long and hard for this money—our lifelong savings. I have devote my life day and night to my ill and dying patients, in order to help the sick and children of our family.

I am enclosing photos of our family, also a lawyer's statement as well as a certificate from the doctor who administered to the last patient I took care of.

Dear Professor Armstrong, I beg you to help us. Donate us the $40,000.00 we were cheated out of our savings.

In return I pray daily to God for your earthly well being and the eternal.

Thanking you in advance with all my heart I send you my regards

[Name withheld]
Flushing, New York

"PAY MY MOTHERS RENT FOR 6 MONTHS"

December 17, 1975

Dear Mr. Armstrong

I am a boy in the 6th grade. My teacher collects autographs. He said if I got a letter handwritten from you I could sell it for $500. I am a very poor boy and a letter from you would give me mony for Christmass presents and to pay my mothers rent for 6 months. I am asking you out of mercy to help me out.

Sincerely Yours

[Name withheld]
Brooklyn, New York

P.S. I always wanted to become a Astronaut.

At the top of this letter Neil's assistant had written "QUACK."

"MY NEAR FUTURE PATENTED INVENTIONS"

October 27, 1976

For Mr. Professor Neil Armstrong
Aerospace Department
798 Rhodes Hall
University of Cincinnati
Ohio

Dear Mr. Professor Neil Armstrong:

I am leaving this letter for you about a future requested meeting of ours to be arranged by you, or your clerical staff, concerning several of my near future patented inventions and their large scale & immense in size & in scope & their unusual far advanced American Scientific benefits to all of America; & as one or more Pentagon officials have said in the past, even they would benefit all of mankind, too.

 I will try to arrange this meeting when you have checked your schedule, Neil, to also introduce you to some Engineering Business Associates of mine, plus to have them be included in on the requested meeting of ours, at either nearby Walnut Hills Firm location, or at your U.C. precise location, or at a 3rd secret location at the E.P.A., to this much long overdue historic meeting of ours Mr. Armstrong.

 Mr. Armstrong, the White House in Washington, D.C., has also been informed, plus the Pentagon itself as mentioned earlier, as to this future Scientific Projects meeting of ours, sir, for later in 11/76 or 12/76 if it cannot be in 11/76 at the mid-way point of the Fall U.C. quarter, sir.

I will leave this short letter with Mrs. Moore for you to reply within the next calendar week, sir.

Sincerely,

[Name withheld]
Cincinnati, Ohio

Reply from Elaine Moore, Neil's secretary

November 9, 1976

[Name and street address withheld]
Cincinnati, OH 45219

Dear [Name withheld]:

Professor Armstrong has asked that I respond to your letter. His schedule is such that he is unable to schedule meetings without prior knowledge of the subject matter. Should you have specific requests to Professor Armstrong, please forward those requests along with sufficient background information to permit a review by his office.

Sincerely,

Elaine Moore
Secretary to Professor Neil A. Armstrong

"10 massive far advanced projects of mine"

November 30, 1976

Mr. Professor Neil Armstrong:

Due to my unusually busy 10 projects schedule, and for several other reasons. Technically I have not been able until now to send this requested follow up additional letter to your 798 Rhodes Hill-U.C. official mailing address. Mrs. Moore mentioned that I should explain the reason for the meetings, on one certain meeting later at the nearby A.M. Kinney Engineering Firm to you Neil in this next letter. I do want you, sir, to reply personally the next time, so as to show your reply to several local business firms & related project officials who later are very interested in discussing my massive related NASA-Police related types of projects with you & I both present, sir. I will be going to Houston, Texas, and the Florida Space Center later next year in 1977, sir, to inspect certain areas for later Space Shuttle operations. I will soon start to use this new NASA related item in my 10 projects' different aspects in the near future, sir. Due to the extreme secrecy of these 10 projects of mine, I will only expect your personal replies to my present and any additional correspondence.

Mr. Armstrong, these 10 massive far advanced projects of mine, sir, are so complex that only near 50 people in the entire U.S. itself can even communicate with my high I-Q-level far advanced Intellectual Level. The Pentagon officials are really amazed these last 2 years as to how I was able to create such amazing new for advanced American Scientific Systems the way I created them. I explained to one of these officials recently, that with *GOD's* eternal help and my never ending research for perfection can outdo any present day 8 year PHD-Scientists anywhere in the U.S., when it relates to nuclear special items, and on NASA technologies, too.

Please note, sir, the meeting at A.M. Kinney Engineering Firm, or the other separate meeting of ours, to discuss these 10 projects of mine, which are Classified as of 1976, sir, should be arranged for the month of 1/77 when it is convenient for you.

Sincerely,

[Name and street address withheld]
Cincinnati, Ohio

Reply from Elaine Moore

December 9, 1976

[Name and street address withheld]
Cincinnati, Ohio

Dear [Name withheld]:

Professor Armstrong has reviewed your material and finds nothing sufficiently specific for him to respond to. He is unable to schedule meetings on such a limited basis.

Sincerely,

Elaine Moore
Secretary to Professor Neil A. Armstrong

"POLICE AND THE F.B.I. HAVE INVESTIGATED YOUR COMPLAINT"

June 19, 1978

To: Prof. Neil Armstrong
From: Det. Carol Allen

U.C. Police and the F.B.I. have investigated your complaint regarding letters from [name withheld]. [Name withheld] does reside at [street address withheld] in Cincinnati and has no criminal record. The F.B.I. has informed me that there has been no federal violation, therefore, no action will be taken at this time. If in the future the letters become threatening in any way, please notify me immediately.

By Direction of Chief E. R. Bridgeman.
[Signed by Col. E. R. Bridgman, University of Cincinnati Police]

"BREAKTHROUGH IN ROCKETS, SPACE SHUTTLES, SPACE STATIONS AND ANYTHING THAT MOVES"

May 8, 1979 [The letter writer dates it "May 8, 1779 + 200"]

Dear Mr. Retired Astronaut,

I am contacting all of the retired Astronauts. Please form a chain letter to coordinate information among you.
 I have made a breakthrough in rockets, space shuttles, space stations and anything that moves. One idea I have is the shaping of moving vehicles front and rear to act like rocket nozzles to create an equal &

opposite reaction thus nullifying wind resistance. I have about 50 ideas related to the formula T = mv.

I want a meeting with every man Jack. I want you to form a corporation "Project *2, Inc." I will toss my ideas in the pot. I will not be in the corporation. I don't want to get involved with high finance & business transactions. I will stay with my one main company. I'm sure you all will have thousands of ideas to toss in the pot.

National defense, Energy, Living Standards, Food, Pollution—our survival are all involved so please move out not to speed 55 MPH,

I will be happy to answer all questions.

Wings on airplanes are equivalent to the Dodo bird. I believe I can correct a B52 to fly at 100,000 feet at 500 MPH. It could rise straight up and land on a dime. It could go into parking orbit at 18,000 MPH, using re-fuel. It could ride in outer space. Do you known any nice old ladies who would like a trip into outer space? How about ships that sail at 200 MPH or a submarine at 100 MPH. I am teaming up with an expert on air flow on projects.

Respectfully,

[Name withheld]
Benjamin Franklin Innovation Company
Memphis, Tennessee

The tagline for the man's company was "WE HAVE JUST BEGIN TO THINK." At the bottom of the letter Neil wrote, "No answer. File under Quack."

"I WOULD LIKE TO TAKE A WOMAN WITH ME"

October 2, 1979

To: Dr. Neil Armstrong, Aerospace Department, University of Cincinnati, Ohio
From: [Name withheld], New Jersey State Prison, Trenton, NJ 08625

Hello Cincinnati! This is Trenton Base. The Eagle has landed!

I want to take one small step for man, from here to Houston and Cape Canaveral. Then one giant leap for mankind from the Launch

Tower at Canaveral through the solar system, an orbit from around outside the Earth.

Is there a Saturn 5 rocket in storage in Florida? Or elsewhere? Any parts?

I would like to take a woman with me for companionship and other reasons—one small step for woman, one giant leap for womankind. Dig?

In six months' time I could be ready for LIFT-OFF. March 1980? I wrote NASA five times since 1978, never heard from them P's!

I also like to go by myself in a way.

I have a good name & reputation. I'm a concerned citizen, patriotic American, nail payer & a voter of myself. Basically life long good person like the rest.

I have given myself a governortorial & Presidential pardon a long time ago on General Principal Grounds.

Maybe a nuclear thing can be installed in the service module? Lots of oxygen tanks, use lightly. Extra fuel, water tank. No food. One tablet every 24 hour period. Books, tape cassets. One or two seats only need to be in the command module. Have name for Apollo CM. I'm in good physical condition, another B Nail Armstrong, Mr. Clean type. I'm ready for this journal, Rare Admiral Odyss, Honararium Member U.S.C.G.

Want to visit me? Magnum opus pro bono public summum bonum.

I won't smoke when I leave here. Don't want newspaper, magazine, TV, other photographer types to do any picture takeing while I'm in training at Houston or elsewhere, or my comrade astronautnesse. I want her to be given the title has a Honoree member U.S.C.G. Act. Captain. It will be alright to film on entering the hatch and so forth.

To have a March 1980 lift-off, I will have to be released real soon to go through intense training. From here I can take a car to Houston. It takes too long to go by horse or bicycle with a 3/80 date set.

I guess I will outdo old Charlie Lindbergh.

I hope to see you soon, old buddy.

[Name withheld]

This man wrote several letters to Armstrong, all from the New Jersey State Prison. The letter reproduced here is the most coherent of them. Neil did not reply to any of them, nor did his secretary.

A RELUCTANT ICON: LETTERS TO NEIL ARMSTRONG

"I CAN'T BELIEVE THEY WOULD REVEAL ANYTHING"

March 10, 1980

Mr. J. Allen Hynek
Center for UFO Studies
Evanston, Illinois 60201

Dear Allen:

As you can see from the above letterhead, I have left the university to join a family enterprise in the oilfield supply business. The associated travel has kept me away, and I'm just now trying to "catch-up" on the stacks of correspondence.

I am not persuaded that there would be any advantage to my attendance at a conference with Mr. Stringfield with or without his witnesses. I can't believe they would reveal anything to me that they would not tell the authorities.

I appreciate your kind comments regarding "The Voyage of Charles Darwin." I'm pleased to be associated with the project.

The Armstrongs send their best to the Hynek family.

Sincerely,

Neil A. Armstrong

It would be inappropriate to present J. Allen Hynek (1910–1986) as a quack or conspiracy theorist, for Neil Armstrong did not consider him such. Certainly Neil did know him as one of the world's leading UFOlogists but also as one of the most credible ones—a trained scientist who in 1935 had earned a PhD in astrophysics at Yerkes Observatory and then taught in the Department of Physics and Astronomy at Ohio State University. Dr. Hynek's research focused on the study of stellar evolution and the identification of spectroscopic binary stars. By the time Neil got to know him, in the early 1970s, Hynek was best known for his UFO research, having been the primary scientific investigator employed by the U.S. Air Force to look into UFO incidents, first with Project Sign (1947–1949), then with Project Grudge (1947–1949), and finally with Project Blue Book (1952–1969). In 1973, Hynek founded in Evanston, Illinois, the Center for UFO Studies, which not only investigated cases of UFO sightings but also created extensive UFO archives that included valuable files from civilian research groups such as the National Investigations Committee on Aerial Phenomena (NICAP),

QUACKS, CONSPIRACY THEORISTS, AND UFOLOGISTS

one of the most active and credible UFO research groups in the U.S. during the 1950s and 1960s.

Armstrong and Hynek, in fact, became relatively close friends. Neil and his family (wife Janet and boys Rick and Mark) had met the Hynek family (wife Miriam and five children) in the summer of 1973 during the two-week-long "eclipse cruise" on board the luxury British ocean liner S.S. Canberra, which left the port of New York City with a destination off the west coast of Africa for prime viewing by its 2,000 passengers of the rare total eclipse of the Sun. The Armstrongs and Hyneks met and became friends because they had been assigned to the same dinner for all their meals during the cruise. (Also on board the eclipse cruise were such "nerd celebrities" as author Isaac Asimov, New York Times science reporter Walter Sullivan, astronauts Alan Shepard and Scott Carpenter, and many more, along with family members.) Neil never came to believe in UFOs, but his friendship and occasional correspondence with Hynek shows that Neil had an open mind on the subject, though he chose to keep that attitude completely private and to himself.

In his March 10, 1980, letter to Hynek, Neil refers to a "Mr. Stringfield." Who he was referring to was Leonard Stringfield (1920–1994), another of the prominent UFOlogists (he served both as director of the organization Civilian Research, Interplanetary Flying Objects [CRIFO] and as public relations director for NICAP)—one who took particular interest in stories about UFOs that "crashed," stories that he published in a monthly newsletter, ORBIT. He eventually wrote a number of books about alleged recoveries of alien spaceships and alien bodies.

UFO enthusiasts who are reading this book may not be happy to learn that there are very few letters extant between Armstrong and Hynek in Purdue University's Neil A. Armstrong papers collection—and that the Hynek family has no possession of any letters between the two men. But it is likely that Neil stayed generally aware of some of Hynek's more high-profile activities. At a symposium in 1973, held in Akron and sponsored by the Mutual UFO Network (MUFON), Hynek attracted headlines by claiming that the number of UFO sightings in the U.S. was much higher than what was reflected in the Project Blue Book statistics. In 1975, he presented a talk on his UFO research at a meeting of the American Institute of Aeronautics Astronautics (AIAA), an organization to which Armstrong actively and proudly belonged. In 1977 Hynek gave a highly publicized talk at the First International UFO Congress, held in Chicago, entitled "What I Really Believe about UFOs." If Neil read the newspaper coverage—and he very well might have—he saw that Hynek very precisely and non-sensationally stated: "I do believe that the UFO phenomenon as a whole is real, but I do not mean necessarily that it's just one thing. We must ask whether the diversity of observed UFOs . . . all spring from the same basic source, as do weather phenomena, which all originate in the atmosphere," or whether they differ "as a rain shower differs from a meteor, which in turn differs from a cosmic-ray shower. . . . We must not ask simply which hypothesis can explain the most facts, but rather which hypothesis can explain the most puzzling facts."[13] In November 1978 Hynek even presented a statement on UFOs before the United Nations General Assembly. The goal of his speech, which went unfulfilled, was to create a centralized U.N. authority on UFOs. In his personal library Neil had copies of three of Hynek's books: The UFO Experience: A Scientific Enquiry *(1972)*, The Edge of Reality: A Progress Report on the Unidentified Flying Objects *(1975, coauthored with Jacques Vallée)*, and The Hynek UFO Report *(1977). In* The Edge of Reality *he proposed a "close encounter" scale designed to better catalog UFO reports; he later served as a consultant to Columbia Pictures and Steven Spielberg for the 1977 Hollywood movie* Close Encounters of the Third Kind, *and even made a cameo appearance near the end of the film. In 2019 the History Channel showed season one of a series based on the case files of Project Blue Book, in which Hynek plays the central character. (He is played by actor Aidan Gillen, who plays Littlefinger in the HBO series* Game of Thrones.*)*

The following letter from Neil to Hynek from August 1984 shows (1) that Dr. Hynek was interested in more than just UFOs, and (2) that Neil and Allen stayed in touch.

"Sending the Hyneks our best"

August 17, 1984

Dr. J. Allen Hynek
Scientific Director
Center for UFO Studies
P.O. Box 1402
Evanston, IL 60204

Dear Allen:

I can certainly understand moving from Chicago to Phoenix! I know you'll enjoy it in the great south west.

The daylight lunar scene, in terms of lighting, is somewhat akin to watching a night football game. The lighted surface causes the pupils to contract, limiting their ability to discern celestial objects. If one were to look through a blackened tube and allow for night adaptation, I'm certain 3rd or 4th magnitude (or smaller) stars could be seen.

We used a 1 power sextant (with sunshade) to take star shots and 2nd magnitude stars successfully.

Janet and the boys join me in sending the Hyneks our best.

Sincerely,

Neil A. Armstrong

"I WANT TO PAY NEIL ARMSTRONG $500,000.00"

June 3, 1981

Professor Neil Armstrong:

I have a trust account from the California Institute of Technology. You must find out the name of the bank that Cal Tech has my money in. I have $2,000,000.00 in this bank account.

I want to pay $250,000.00 for each FBI agent at three agents to get me out of Greystone Park and take me to the trust account to be paid. Private I investigators are my second choice.

I want to pay Neil Armstrong $500,000.00 for hiring the three FBI agents to have taken me out of Greystone Park and to the trust account. Neil Armstrong will be paid the money at the bank when I arrive.

The best time to come up is between 1 and 3 PM or 6 and 8 PM. A Brinks truck is a nice idea for transportation.

Sincerely,

[Name withheld]
Morris Plains, New Jersey

Neil told his secretary not to reply to this letter.

A RELUCTANT ICON: LETTERS TO NEIL ARMSTRONG

"I SAW TWO UFOS"

June 17, 1981

Astronaut Neil Armstrong:

While scouting downtown L.A. one day President Nixon was in office I saw two UFOs. As I was unacquainted @ Project Bluebook's address I sent the above information to the A.F. records H.Q.

They appeared thusly [the letter writer draws two dots, labeling them one silver and one black] and traveled in straight lines in a 1½ degree climb. They appeared to be observation craft and nothing more. Naturally I did not send up *Red Secret*. The main reason was that, as it rose from the water tank in which it rests submerged, it may have flooded the courtyard of the Alpine Terrace Castle, across the private access road, and perhaps powered down the hillside of the Castle's site, sir.

Naturally, following this phenomenon, I was greatly relieved to see a WACS plane fly inland from the ocean as it emerged from a group of cumulous clouds. I left the roof of the 10 story building and took the elevator to the street level and drove home to Alhambra.

Incidentally, how is Buzz? He looked great at the Balsa Mall the other day, and is a brilliant speaker. He didn't mention that he splashed YAK 9s but I'll bet he didn't splash one Sturmovik. He said you and Mike Collins were tops and Mike sent my black address book back with $1.00 still in the spiral.

Well, Edwin expounded on the Space Shuttle and held us in the palm of his hand as he explained that it is:

1. The 1st U.S. space venture (manned) since the Apollo spaceships.
2. You, he, and Mike were a team, re: America's Moon.
3. He personally saw the Space Shuttle touch down successfully.

Peace be with you and yours, and hopefully President Reagan will find amusement in *6 Years From Earth*.

What a Life Secret, I had 5 Merit Badges: First Aid, Cooking, Pathfinding, Art, and Swimming. And when in Army R.O.T.C., I was a 2nd Lt., then a 1st Lt., both lifetime commissions; additionally I was Rifle Team Captain plus Drill Team Captain. And in the U.S.A.F. I was:

Rank A/B-pay grade, E1, A/3C-pay grade-E3, and A/1C-pay grade E4.

You, Sir, I am also enclosing a copy of my latest resume, plus a copy of Golden Spider.

Best regards from a friend,

[Name withheld]
Phoenix, Arizona

The Yak-9 (Yakovlev) was a single-engine World War II–era Soviet fighter aircraft used by the North Korean Air Force during the Korean War. The Sturmovik was the Ilyushin Il-2, a ground attack aircraft produced by the Soviet Union during World War II. During the Korean War, the Korean People's Air Force operated some fifty Il-10s, a later version of the aircraft. Buzz Aldrin served as a jet fighter pilot during the Korean War, flying 66 combat missions in F-86 Sabres during which he shot down two Mikoyan-Gurevich MiG-15 aircraft.

"IT IS YOUR DUTY TO CONTROL WAR"

November 11, 1983

Dear Armstrong,

Can you remember my name?

I consider you a very important person, an international genius, linked with world peace and involved in nuclear warfare.

You will be surprised to know that a group of people tried to homicide me at my own country where as my name was recommended for future president of India.

King Fagd, UAE Syrian country, is my well wisher. While my life was at stake by enemy attack, noble King Fahd tried to save my life different ways. Her daughter Zulie was interested about me.

I noticed that Syria is attacked by Israel dated Nov. 21, 1983. Israeli war planes targeted towards Syrian held mountains.

I know that Israel is energized by U.S.A. and Netajee is alive living in Israel. My assessment that Netajee developed Israel.

You are requested to convince Israel American commanders and Israel prime minister not to attack my well wisher country Syria.

Your influence your government and force Israel to be friendly with Syria.

As you people given much help to Israel, now it is your duty to control war.

I hope I will be able to be vice-president of India in near future.

I shall find time to see you at NASA, USA, to learn how to take defensive action against enemy attack.

Yours sincerely,

[Name withheld]
Bombay, India

"NO EVIDENCE HAS BEEN FOUND"

December 4, 1985

Police Sergeant Malcolm Stewart
Scottish Police College
Tull. Allan
Kincardine
Fife
Scotland

Dear Malcolm,

I just returned from meetings in Morocco and Brazil to find your letter. As you are almost at your "deadline," I am responding immediately, although with very little information.

On all of the manned and unmanned flights to the moon, no evidence has been found which supports the existence of, or visits by, any living creatures. No water, hydrogen, or carbon has been found except at the molecular level.

There is increasing interest in returning to the moon in the future; so perhaps I will be proven wrong! There is hope that water (ice) will be found in the craters near the lunar poles.

I hope you received the Apollo slides and tapes I sent you a year or two ago. I got them from the Smithsonian Institution in Washington.

QUACKS, CONSPIRACY THEORISTS, AND UFOLOGISTS

I hope I can accept your invitation to the Old Course sometime soon. You can bet I'll not miss the opportunity if it arises.

All the best,

Neil A. Armstrong

In August 1976 Armstrong accompanied a Scottish regiment—called the Black Watch and the Royal Highland Fusiliers—on an expedition into the vast Cueva de los Tayos (Caves of the Oil Birds), located on the eastern slopes of the Andes Mountains in a remote part of Ecuador. The caves, first discovered in 1969 by a Hungarian-Argentinian adventurer, János Juan Móricz (1923–1991), reportedly contained mounds of gold, some unusual sculptures, and chunks of metal with some manner of writing apparently engraved into them. Because of Neil's Scottish ancestry, and the fact that the British side of the project was mainly Scottish, he was invited by the expedition's project director, Edinburgh engineer Stanley Hall, to act as the exploration's honorary chair. Neil accepted but, uncharacteristically, did not know much about the background to the whole affair. Chiefly, he was not aware that the premise for the "Expedición Los Tayos" came from controversial Swiss author Erich von Däniken's 1972 book, The Gold of the Gods, *the follow-up publication to his sensational best seller from 1968* Chariots of the Gods? Unsolved Mysteries of the Past *(and the subsequent West German film documentary of 1970 of that name, which was one of the highest grossing films of the 1970s and an Academy Award nominee as Best Documentary Film). In these works von Däniken theorized that extraterrestrials not only had come to Earth several centuries earlier but had deeply impacted the course of early human life but actually laid the basis for the origins of civilization. In* The Gold of the Gods, *von Däniken described what he claimed was his own exploration into the Cueva de los Tayos, asserting that he had found considerable archaeological evidence of an extraterrestrial presence, including certain doorways in the cave that were too square to have been made naturally, lifelike statues of humans and animals, furniture made of plastic and gold, and thousands of metal and crystal plates precisely etched in an alien language and recording 250,000 years of history. Armstrong later related for* First Man *that he had not read any of von Däniken's books before his trip to Ecuador, did not know of any connection that the Swiss writer might have with the caves—nor did he know that Stanley Hall, the expedition director, was himself a theorist about ancient aliens. But not long after Neil boarded a Royal Air Force plane* for Quito (in the company of Hall and select volunteers of the Black Watch and the Royal Highland Fusiliers) he quickly learned about all of it. At that point there was nothing for him to do but make the expedition. The man who innocently told him the most about what was going on was police sergeant Malcolm Stewart, who taught at the Scottish Police College in Kincardine, on the Firth of Forth, Country Fife, Scotland, some forty-five miles southwest of the historic golfing town of St. Andrews.

Neil did not become a convert to anyone's theories about "ancient astronauts," though in the aftermath of the 1976 expedition, von Däniken (who did not go on the expedition) and others tried to turn him into a supporter of von Däniken's sensational notion that extraterrestrial beings had visited Earth in the remote past, leaving various archaeological traces of their civilization-building activities. But Neil would have no part of that, though he chose to make no public statements whatsoever, pro or con, about any of the findings or hypotheses purported by von Däniken and the others. Neil also declined at least two invitations from von Däniken to join him in future expeditions to the caves in Ecuador. In private Neil did tell friends and family that what the 1976 expedition found in the caves of Los Tayos were "natural formations." When this word got out, von Däniken sent Armstrong a two-page letter, dated February 1977, in which he told Neil that the expedition he had been on "cannot possibly have been to my cave." Neil made no further replies to mail from von Däniken.

For the next ten years, into the mid-1980s, Neil replied to occasional letters sent to him by expeditioners Stanley Hall and Malcolm Stewart, always with friendliness and cordiality.

In 2019 a detailed examination of the controversial expeditions to the Tayos Cave complex in Ecuador was published by journalist, explorer, and filmmaker Alex Chionetti entitled Mysteries of the Tayos Caves: The Lost Civilizations Where the Andes Meet the Amazon. *Chionetti's book discusses not only Stanley Hall's quest with Neil Armstrong but also expeditions from the 1960s and 1970s involving the Mormon Church's search for lost tablets. At author Chionetti's request, I provided him copies of some of Armstrong's correspondence with Stanley Hall and Malcolm Stewart from the years 1976 to 1985.*

QUACKS, CONSPIRACY THEORISTS, AND UFOLOGISTS

"SIGHTINGS FROM ASTRONAUTS"

[Undated, circa 1999]

Neil Armstrong
P.O. Box 436
Lebanon, Ohio 45036
USA

Dear Neil:

My name is [name withheld]. I am writing to you about an essay that I had written and a letter I had written to former president Jimmy Carter. The essay was about UFOs and the so-called sightings from astronauts while in space. As to why I had written to Jimmy Carter, well, I read a book by a guy named Timothy Good and guess what the book was, yep UFOs. And it had a piece in it about how Carter didn't get reelected because he was going to let loose all the information about UFO's public so you can imagine what most likely happened next. I know you and Jimmy Carter have nothing in common and you are not a politician. The fact still remains that astronauts have been linked to UFO sightings in space. All I want to know is your opinion not some desk jockey in Washington. My address is [street address withheld] Victoria, Australia. I do not know how much of this is true but can you please let me know.

Sincerely,

[Name withheld]

To his letter to Armstrong the letter writer attached his letter to Jimmy Carter, along with the essay he had written about UFOs. The title of his essay was "I Believe," a phrase that he explains was made famous by the popular TV series The X Files. The final paragraph of his essay read: "To deny any information on UFO's in my eyes is an insult to the intelligence of anyone who wants to know about UFO's. On the verge of the 21st century some governments treat the population like we are still in the 1950's with the War of the Worlds era still fresh in peoples minds. The Governments cannot throw the blanket over the eyes of the World forever. As the writing says on the poster in Agent Mulder's office in the 'X-Files,' 'I Believe.'"

A RELUCTANT ICON: LETTERS TO NEIL ARMSTRONG

The book by Timothy Good mentioned in this letter was Beyond Top Secret: The Worldwide UFO Security Threat, *published in 1989. (Good went on to publish a number of books on UFOs.)*

Neil did not reply to this letter, but his administrative aide, Vivian White, did send out the generic reply that had been going out to everyone who wrote to Neil with questions about UFOs.

Incidentally, the letter writer's reference to Jimmy Carter was not baseless. On September 18, 1973, while Governor of Georgia, Carter filed a report with the National Investigations Committee on Aerial Phenomena (NICAP), claiming he had seen a UFO back in October 1969. During the presidential campaign of 1976 (in which Carter defeated Gerald Ford for the presidency), Carter admitted that he had seen a UFO. It happened while he was waiting outside for a meeting of the Lion's Club in Leary, Georgia. At about 7:30 p.m. the future president spotted in the sky what he called "the darndest thing I've ever seen." Besides Carter, ten to twelve other people witnessed the event, describing the object as "very bright with changing colors and about the size of the moon" Carter reported that "the object hovered about 30 degrees above the horizon and moved in toward the earth and away before disappearing into the distance." He later told a reporter that, after the experience, he vowed never again to ridicule anyone who claimed to have seen a UFO. During the 1976 presidential campaign, Carter promised that, if elected, he would encourage the government to release to the public "every piece of information" about UFOs available. After entering the White House, Carter backed away from his pledge, saying that the release of some information could have "defense implications" and possibly pose a threat to national security. Believers in UFOs, like this letter writer from Australia, took Carter's action to indicate that the U.S. government was, in fact, covering up the reality of UFOs. [14]

"I KNOW ... THAT YOU DID NOT LAND ON THE MOON OR EVEN ORBIT IT"

September 6, 1999

Mr. Armstrong,

Enclosed you will find excerpts of classified videotapes that I serendipitously received; a thirteen minute sample, including a brief introduction and explanation. It shows yourself, Michael Collins, and Edwin Aldrin Jr., "faking" part of the Apollo 11 missions, specifically being "half way to the moon" or "130,000 miles out" and "177,000 miles out." The original tapes in my possession are well over 90 minutes in length and are logged "July 18, 1969" and "July 19, 1969" when, according to your own flight plan, you were supposed to be approaching lunar orbit.

These tapes show the illusion that your crew created with the circular window of the darkened spacecraft in low earth orbit as if it were the diameter of the earth at a distance, as well as very clear private conversations with a third party (not Houston) where the transmissions were being sent and recorded for "playback later." NBC, where I worked for two years previously in news, has copies, yet will not air them or copy them further without my permission unless I am incapacitated or deceased.

I am the persistent young man who found out where you resided twice and sought you in earnest just to ask you, face to face, off the record, if the allegations brought to me by a man who had high security clearances at Rocketdyne were true. I first must ask your forgiveness for any inappropriate or immature conduct on my part as I first took on this project five years ago. I sometimes have zeal without forethought. Also, though, I know now, not then, that you did not land on the moon or even orbit it.

When I first saw the tape and understood what it meant, I was filled with sorrow and wept. Even though I had been working on the project for five years, I did not believe it until I saw it with my own eyes. I also believe that you and your colleagues were hard pressed, perhaps with lethal threats on your family, to participate. If anyone from the crews resigned, especially the first crew, too many questions would have been asked. It is my own conviction, one which I will share, that these twenty men, and yourself, were not responsible.

I have enclosed a photograph of my son to let you know that I have a family too. In the past two months his life and mine have been put in danger as well because of these tapes. I was apprehended just outside the front door of CNN in Atlanta by three men identifying themselves as police officers before I could show them to the assignment editor inside. I had the good graces of God to escaped from where they detained me. My attorney has informed me that there is no record of my arrest or of the videotape copies they confiscated.

On the 25th anniversary in 1994, as you held back tears and challenged the young students touring the White House to "perhaps, someday, remove one of *truth's protective layers*," I saw hope in your eyes. A couple of months ago, when you issued another brief statement, I saw resignation. Am I right? Can you teach an old dog new tricks?

As you near seventy, and your associates depart this life at that very age, I wonder. Is anyone fearful of a deathbed confession from the dwindling number of men who know the truth? How many more, your friends, and possibly me, must die for the dark decision of a few men thirty years ago? The Bible says, "*Satan was a murderer from the beginning, not holding to the truth, for there is no truth in him. When he lies, he speaks his native language, for he is a liar and the father of lies*" (John 8:44). I know now that the reason you never give interviews is so you won't have to lie.

I believe in you with all my heart. I have studied your life. I know you from the symbolic nature of this event, and even deeper, the symbolic nature of this lie. There is only one life for everyone. Do you really think it will matter to all those in the world, past, present, and future, who will join us all in the eternal life to come that we can not escape from, whether or not we went to the moon? The important question is for the living. What will be for the lasting good of the world and our country; to darken the planet and its citizens darker and darker each year with perpetual prideful stubbornness, unwilling to admit wrong, or to purge our consciences, and avenge the innocent murdered, and throw off the yoke of slavery to sin and to an arrogant government that needs to have the gangrene cut off.

I believe that you and your brothers must break your vow of silence. You are in danger until you do so. Once this is brought out into the light the roaches will run for the shadows. At that time you and they, and your families, will be untouchable because to harm you then would draw enormous suspicion. The Bible also says, "*Whoever exalts himself will be humbled, and whoever humbles himself will be exalted*" (Matthew 23:12).

Now is the time for the true hero to come forward, for the true American patriot to lead the way, like Washington, who was willing to tell it like it was and let the chips fall as they may, trusting God that the betterment of the country would always come from the facts. I am waiting to hear from you before this footage airs worldwide. Please think of my two year old son. I am willing to interview you on your terms, including not releasing the footage or your interview until a time and place that you decide.

My Church is the True Church written about in the New Testament. Their Cincinnati telephone number is [withheld]. Call to find the current meeting location. I encourage you to attend a Sunday service anonymously and check it out, and to let us serve you, even provide for you a safe place for you and your wife anywhere in the world, should you need it. Not I, God, truly awaits your answer. "For whoever wants to save his life will lose it, but whoever loses his life for me will find it. What good will it be for a man if he gains the whole world yet forfeits his soul."

Yours sincerely,

[Name and address withheld]

Reply from Vivian White

September 13, 1999

[Name and address withheld]

Mr. [Name withheld]:

Mr. Armstrong asked me to notify you that your accusation was received. He notes that you are calling him and the entire U.S. Government accomplices to fraud.

He asked me to tell you that you are just simply wrong. He has no objections to your releasing the tapes to anyone you wish.

Sincerely,

Vivian White
Administrative Aide

"Kook has been stalking me for some years"

September 13, 1999

Mr. Daniel S. Goldin
Administrator
National Aeronautics and Space Administration
Washington, D.C.

A RELUCTANT ICON: LETTERS TO NEIL ARMSTRONG

Dear Dan:

Enclosed please find a copy of a letter received recently and a copy of the response of my assistant.

This kook has been stalking me for some years. He may be known to your security people. I would have written to them directly, but didn't know a proper name or office address.

Sincerely,

Neil A. Armstrong

"Proves that the fire was set deliberately"

September 16, 1999

Mr. Armstrong,

I don't know what to believe. Surely you're not saying that the federal government has never lied or misled the public? As for you, only you know your motives. Are you trying to protect me? Please explain what was going on on the videotape that I sent you. This will help pursued me.

I know you are wiser and smarter than me, nevertheless, the facts are presenting themselves that cannot be refuted by denial alone. What is really going on? I have spoken with Scott Grissom on several occasions and he has told me plainly that his father was murdered. I know you know him and the fact that he is no dummy. He has also informed me that he has tangible proof, removed from the spacecraft, that proves that the fire was set deliberately. I will be doing a film about this too.

We all know about the lemon incident. What would Gus have done if he were asked to do something he refused to do? Surely this was someone's concern. Are you willing to let his murderers and his death escape justice? Are you willing to go to the grave with this and untruths about history forever published in encyclopedias? I don't know what to say. I am not calling you a liar, or am I? We are all liars. Everyone. What happened was us not "them." If I am deceived, I beg you to pursued me.

Enclosed is a new book on the subject. This issue will not die. Perhaps it is because there is a battle going on for the truth. The question is, what is right? Everyone who does something wrong justifies it to himself, that does not make it right. What is right is worth dying for. We're going to die anyway. Even if the government's destruction would come from the truth, then it is not worthy to stand, and its betterment would inevitably follow.

Your servant,

[Name and address withheld]

Neil did not reply to the man's second letter, for he had already seen far too much of the self-proclaimed "investigative reporter." In his first letter, the man asked Neil for his "forgiveness for any inappropriate or immature conduct on my part" a few years earlier. What the man was referring to is the following. In 1996 the man had entered uninvited into the Armstrongs' suburban Cincinnati home. In an interview for First Man, *Neil's second wife, Carol Held Knight Armstrong, related what happened: "Neil was at the office. This guy knocked at the door and there was a big dog with him, and he had a package. I opened the outside door while leaving the screen door shut, and the man said, 'Is Neil here?' I said, 'No, he's not. May I help you?' He opened the screen door and just walked in, bringing along his dog. He said, 'I want him to sign this,' and I said, 'Neil doesn't sign things anymore.' 'He'll sign this,' he uttered, and then he left. It sort of hit me three minutes later. All of a sudden I felt shaky." In the following weeks, the interloper started putting letters and other things in the Armstrongs' mailbox. Some of the materials had religious overtones and most were about the Moon landing being faked. The local police department responded, "It's probably nothing, but why don't you just bring the tapes and letters and we'll take a look at them," until a call to the ABC TV station in [city withheld], where the man said he worked, revealed that he had never worked there but instead was an independent filmmaker who operated a business called [name withheld]. A few weeks later, Carol received a phone call from her neighbor: "Carol, there's this car parked out here and it's been out here for a long time." When the neighbor went out to investigate, she saw a lot of camera equipment in the back seat. The siege continued for three days, culminating in a car chase involving the Armstrongs, the man, and the police.* [15]

Then in September 1999 Armstrong received the man's disturbing letter. Later that year Fox television network broadcast a "documentary" entitled Conspiracy Theory: Did We Land on the Moon? *that was based largely on the low-budget commercial video the man had produced called* A Funny Thing Happened on the Way to the Moon, *excepts of which he had mailed to Neil with his letter.*

A RELUCTANT ICON: LETTERS TO NEIL ARMSTRONG

The Fox program speculated, based mostly on the man's video, that the Moon landings were an ingenious ploy of the U.S. government to win the Cold War and stimulate the collapse of Soviet communism by forcing the Kremlin into investing massive sums of money on its own lunar program, thereby ruining the Russian economy and provoking the internal downfall of the government. Little wonder that Neil did not continue any sort of correspondence with him, after having Vivian White respond with a curt note to the man's first letter.

But, as we see here, the man did not go away. He sent a follow-up letter that was just as disturbing as the first. In the following two years, he followed Neil to different meetings and on more than one occasion parked directly outside his home until the police intervened. At the annual meeting of EDO Corporation stockholders in New York City in 2001, a board on which Neil had served for several years, the man showed up with a video-camera-carrying assistant. EDO president James Smith recalled the scene: "This guy shows up with a Bible and shouts out, 'Neil Armstrong, will you swear on this Bible that you went to the Moon?' Well, the audience immediately started booing the intruder very loud, but he went right on, 'Everybody else in the world knows you didn't, so why don't you just admit it?!' It quickly turned into a kind of pushy-shovy thing, so I and a few other men got the guy out of there. Subsequent to that, we never had a meeting where we didn't hire special security."[16]

"Had I the opportunity to run that episode over in my life," Armstrong commented for First Man, "I wouldn't have allowed my company people to usher me out of the room. I would have just talked to the crowd and said, 'This person believes that the United States government has committed fraud on all of you, and simultaneously he wants to exercise his right protected by the U.S. government to state his opinions freely to you.'"[17]

A few months after the EDO meeting, on September 9, 2002, the man, his Bible in hand, confronted Buzz Aldrin outside of a Beverly Hills hotel. A resident of the Los Angeles area, Buzz had arrived at the hotel thinking he was to be interviewed by a Japanese educational television network. At first Aldrin, his stepdaughter in tow, tried to answer the man's questions, then did his best to get away from him. But the insistent conspiracy theorist dogged him out of the hotel, directing his assistant to keep the camera running, while shouting at Buzz, "You are a coward and a liar." Harassed to the point of complete exasperation, Aldrin, then seventy-two years old and 160 pounds, decked the thirty-seven-year-old 250-pounder with a quick left hook to the jaw. The man filed a police report but, after watching the accuser's own tape of the incident, the L.A. County District Attorney rather forcefully declined to file charges. As the self-proclaimed "victim," the man later told reporters, "If I walked on the Moon and some guy said swear on a Bible, I'd swear on a stack of Bibles."[18]

The man continued to harass the Apollo astronauts for the purpose of his "filmmaking." In 2004 he released Astronauts Gone Wild: Investigation into the Authenticity of the Moon Landings, *a title that was wordplay on the* Girls Gone Wild *video series*. By being misleading about who he was (his company's name made it look like he was part of ABC News) and what he was doing and why, he was able to arrange interviews with Apollo astronauts Alan Bean, Eugene Cernan, and Edgar Mitchell. When, quickly, each astronaut detected the pretense, the man asked them to "swear and affirm, under penalty of eternal damnation, perjury and treason" that they had really gone to the Moon. Ed Mitchell would later say the following about his encounter: "[The man] faked his way into my home with false History Channel credentials for an interview. After about 3–4 minutes, he popped the Bible question. Realizing who he was, I maintained my cool enough to swear on his bible, then ended the interview and tossed him out of the house, with a boot in his rear."[19] For his video, the man also ambushed Michael Collins, Alfred Worden, Bill Anders, and John Young; he had not been able to arrange formal interviews with any of them but instead accosted them while they were walking through airports or at other public events.

When Neil was asked to swear on the man's Bible at the EDO meeting, he refused to do it, saying, "Mr. [name withheld], you do not deserve answers. Knowing you, that is probably a fake Bible."[20]

> "The only thing more difficult to achieve than the lunar flights would be to successfully fake them"

Following Fox's airing of Conspiracy Theory: Did We Land on the Moon? *in late 1999, Neil Armstrong, in reply to all inquiries about any sort of Moon landing hoax, had his aide Vivian White send out the following letter.*

Dear _____:

I am responding on behalf of Mr. Armstrong to your recent letter regarding the reality of the Apollo program flights.

The flights are undisputed in the scientific and technical worlds. All of the reputable scientific societies affirm the flights and their results.

The crews were observed to enter their spacecraft in Florida and observed to be recovered in the Pacific Ocean. The flights were tracked by radars in a number of countries throughout their flight to the Moon and return. The crew sent television pictures of the voyage including flying over the lunar landscape and on the surface, pictures of lunar scenes previously unknown and now confirmed. The crews returned samples from the lunar surface including some minerals never found on Earth.

Mr. Armstrong believes that the only thing more difficult to achieve than the lunar flights would be to successfully fake them.

Mr. Armstrong accepts that individuals may believe whatever they wish. He was, however, substantially offended by the FOX program's implication that his fellow Apollo crewmen were possible accomplices in the murder of his very good friends, Grissom, White, and Chaffee, and he has indicated his displeasure to FOX.

We appreciate your inquiry and send best wishes.

Sincerely,

Vivian White
Administrative Aide

Fox television network's sensationalistic program had parroted the same uninformed arguments about Apollo that had been around for more than two decades—that is, that the American flag planted by Apollo 11 appears to be waving in a place where there can be no wind; that there are no stars in any of the photographs taken on the lunar surface; that the photographs taken by the Apollo astronauts are simply "too good" to be true; that the 200-degree-plus Moon surface temperatures would have baked the camera film; that the force of the lunar module's descent engine should have created a crater under the module; that no one can travel safely through the "killer radiation" of the Van Allen belts; and more. All of this "evidence" for the Moon landings having been faked had been completely explained by different physicists, engineers, historians, and other experts on the space environment and space history. Still, latest new wave brought an even more active round of complete explanation, notably by Jay Windley, who developed a website, Moon Base Clavius, dedicated to debunking all the Moon hoax allegations. Another persuasive explanation came from Jim McDade, who wrote a long feature story for the Birmingham News *(Alabama) entitled "Lunar Lunacy: Shooting Down Theories That Apollo Moon Landing Was Hoax Conspiracy" dated April 1, 2001. McDade's article characterized the hoax videos as "full of falsehoods, innuendo, strident accusations, half-truths, flawed logic and premature conclusions. . . . [The* conspiracy theorists have] misinterpreted things that are immediately obvious to anyone who has extensively read Apollo history and documentation or anyone who has ever been inside an Apollo Command Module or accurate mockup."

One the first waves of conspiracy theories about the Moon landing had come with the release of the 1978 Hollywood film Capricorn One *(directed by Peter Hyams and starring Elliott Gould, James Brolin, Brenda Vaccaro, and Sam Waterston). The conspiracy in the movie was actually not about the Moon landings but about a fictional first manned mission to Mars. In the imaginative tale, NASA attempted to cover for a highly defective spacecraft by forcing its astronauts before cameras in a desert film studio to act out the journey and trick the world into believing they made the trip. Though a mediocre movie, its notion of a government conspiracy fueled Moon landing conspiracies and continued to do it for years thereafter.*

"HIGHER INTELLIGENCE OFF THE EARTH HAS AN INTEREST IN OUR PLANET"

June 8, 2004

Dear Mr. Armstrong:

I am writing to inform you that our Space Science Division has recently received an incoming transmission that originated from a source in deep space.

The information that we have received indicates that a superior intelligence exists in the universe that has developed the capability to monitor planets that support human life.

They have the capability to visit planets without being detected. Our technology is not sophisticated enough to detect their presence. This highly advanced intelligence visits life supporting planets frequently without being discovered.

The information that we have received authenticates that our planet and our evolution has been watched and observed down through the ages. This superior intelligence has been observing us.

Conclusive evidence now substantiates the facts. There is intelligent life in the universe that visits planets that support human life, and monitors their progress.

We have documented the information that we have received and have compiled it into a book. We have named the book, the Book of the Universe.

The Book of the Universe contains information that is not available here on Earth. The information it contains came to us from "Beyond the Earth."

A selected group of government officials at the federal and State level, and a few well-known persons are receiving copies of the Book of the Universe in order to make this information known.

The Book of the Universe discloses substantial information about a superior intelligence in the universe who has an interest in planets that contain human life.

This superior intelligence has an interest in more advanced civilizations. And seeks them out.

Higher intelligence off the Earth has an interest in our planet. Information we have received confirms this. The convincing facts have been documented in the Book of the Universe. The information we have received is beyond belief. It defies human comprehension.

The Book of the Universe will reach you soon. Study it well and you will be enlightened beyond your wildest dreams.

Very Sincerely, Respectfully,

[Name withheld]
Project Coordinator
Space Science Laboratory
Phoenix, Arizona

It seems that this gentleman ran the Space Science Laboratory out of his apartment in Phoenix.

"THE ASTRONAUTS WERE 60 FEET TALL"

April 26, 2008

Dear Mr. Neil Armstrong,

I am a researcher from India. I wish to convey some of my research findings. I hope that these findings will be useful information and that will excite you.

As it happens to any new discoveries, my findings would be hard to believe. But I have backed up the findings with solid proofs from Apollo mission photographs.

The main finding is that all Earth-based objects including the astronauts had expanded by a factor of 6 to 10 when they were on the Moon. The astronauts were 60 feet tall. The 60 feet tall astronauts, who were unaware of the expansion occurred to them, tried to measure distances and angles on the Moon. In all cases the astronauts had underestimated the distances and angles.

During Apollo 11, you landed south-west of Little West Crater. The Little West Crater was apparently so small that you (yourself and Aldrin) have never identified it. At the end of the EVA, you have visited the Little West Crater But, the 180 meter diameter crater had appeared as a 3-meter crater and it had been mistaken as the Little West Crater.

The events which had misled NASA scientists to mistake the West Crater as the Little West Crater had been brought as a book, which is available at Amazon.

I don't know how the NASA scientists had misread the phenomenon. However if you wish to further and if you could find some time, I can meet you and explain the events.

Yours sincerely,

Rajasekar Balasundarum
Chennai, Tamil
India

Mr. Balasundarum wrote two short books on his Apollo "discoveries": The Apollo Observation NASA Failed to Observe: Earth-Based Objects 10 Times on the Moon *(December 2007) and* The Apollo Observation NASA Failed to Observe II—How the Moon Deceived Neil Armstrong *(January 2008). The publisher of both books, Novel Corporation (which seems to have been created by Mr. Balasundarum himself), described the contents of the second book as follows: "During the Apollo 11 mission Neil Armstrong had seen a crater the size of a football field just before landing on the Moon and the spacecraft was about land on that crater. Neil Armstrong had avoided that crater and safely landed away from it. The crater was the 180 meter diameter West Crater. After setting foot on the Moon, he had expected to see the rim of that large crater. But he could not find any such crater in the same direction where it was supposed to be. Instead he had seen a small crater in that direction. He had visited the small crater and had found its diameter to be about 30 meters. Before landing the astronauts had not seen any crater other than the football field sized crater. Then how did the small crater happen to come in between? This book purports to explain this with a well-researched and proven theory. The fact is that the crater he had seen before landing and the crater he had visited were one and the same. As it happens to all Earth-based objects on reaching the Moon, the dimensions of the Apollo 11 astronauts, their spacecraft and all the other objects they had carried along with them had expanded by a factor of ten. The 60 foot tall Neil Armstrong, who did not know that he had expanded, had underestimated the crater, 180 meters in diameter, as a mere 30 meter diameter crater. How it is the whole world missed this fantastic phenomenon?"* Obviously Mr. Balasundarum used very much the same text in his letter to Armstrong.

4
FELLOW ASTRONAUTS AND THE WORLD OF FLIGHT

In August 1971 Neil resigned his NASA post for a teaching post at the University of Cincinnati. By that time he had not been an astronaut for over a year, having accepted a request from NASA Administrator Thomas O. Paine in April 1970 (the first letter in this chapter) to serve as deputy associate administrator for aeronautics in the Office of Advanced Research and Technology (OART) at NASA Headquarters.

Neil would have preferred to remain in the Astronaut Corps and continue to fly space missions. But by the time he received Administrator Paine's letter, he knew that would not be allowed to do either. As the First Man on the Moon, Neil was too important, too powerful a symbol, to risk ever again on a space flight. Faced with that reality, he concluded that the NASA aeronautics job was something he could do.

But it did not take long for him to realize that an office job in Washington, D.C., was not to his liking. His main frustration was not with his aeronautics job per se but with the ongoing "requests" from NASA, Congress, and the White House for "appearances on demand," which Neil came to find "a real burden."[21] Also, many an evening had to be spent on the Washington dinner party circuit, which was not easy financially—in terms of clothes for himself and his wife—as he was still on the government pay scale.

So when the president of the University of Cincinnati wrote him, for a second time, with an offer of a professorship in the department of aerospace engineering, Neil accepted. Curiously, NASA offered little resistance to Armstrong's departure. The new Nixon-appointed NASA Administrator,

A RELUCTANT ICON: LETTERS TO NEIL ARMSTRONG

Dr. James C. Fletcher, issued the following statement: "It is with special regret that I accept Neil Armstrong's resignation, and I wish him well in his new duties at the University of Cincinnati. His contribution to the National Aeronautics and Space Administration went far beyond his role as an astronaut and as a commander of the first Moon landing. He joined the Agency as a research test pilot, and he leaves it as Deputy Administrator for Aeronautics. In all his duties he has served with distinction and dedication."[22] Neil's time in civilian government service totaled sixteen and a half years, having begun with the National Advisory Committee for Aeronautics (NACA) in 1956. Fletcher indicated that Armstrong would serve as a part-time NASA consultant, something that Neil personally did not plan on doing, or, in the end, actually do.

Perhaps surprisingly, Neil did not maintain close associations with NASA or with his former colleagues at NASA, not even with his fellow astronauts. This chapter exhibits samples of the contacts and friendships he did maintain. Of all the astronauts he had known and worked with, the men he stayed in touch with the most were Jim Lovell and fellow Purdue alumnus Gene Cernan, but he did not communicate very frequently even with them—and when he did it was mostly in his later years. He also stayed friendly with Mike Collins. Curiously, in Purdue University's Neil A. Armstrong papers collection, there is not a single letter from, or to, Buzz Aldrin, only printed copies of a couple of emails from 2009 concerning the fortieth anniversary of Apollo 11. Surely they did communicate some, but it was primarily at anniversary events for Apollo—those that Neil chose to attend.

It is interesting to see in the limited correspondence that Neil did have with former fellow astronauts how often they were asking Neil to do them a favor, or participate in some event, and how often Neil turned them down. Of course, he did do them a lot of favors, but the principles he applied to determine whether he would get involved in an event or activity seem to have applied to his dealing with other astronauts just as much as they applied to everyone else.

In essence, Armstrong would move on to the new challenges of university teaching and, after he was finished with that, in 1979, to those arising from his involvement on the boards of several different American corporations. He refused to live in the past and made many new friends and contacts. For most people around world, he would always be remembered as the iconic First Man on the Moon. But Neil was not in the least interested in resting on his laurels, which he thought were greatly exaggerated in the popular mind, anyway. He wanted to continue to achieve, to contribute, and to live as normal a life as possible, away from the glare of celebrity. He did his absolute best to do that, even though the world mostly would not let him.

"WE WOULD LIKE YOU TO CONSIDER TWO POTENTIAL TOP JOBS IN HEADQUARTERS"

April 6, 1970

Dear Neil:

This is a critical period for U.S. aeronautics, and George Low and I are considering two matters which require early decisions. First, as you know, NASA and the Department of Transportation are now engaged, at the direction of the Senate Space Committee, in a joint study of future U.S. air transportation system needs and the resulting requirements for the new NASA R&D in civil aeronautics. This is a very important activity for our future, but we are not satisfied that sufficient progress is being made by this study group. Second, we recognize the need for NASA to put more brains and muscle into its Aeronautics Program, to include a possible reorganization of this work within NASA to give Aeronautics more top level attention and visibility. We would be interested in your views on these issues and on the part you yourself might play in helping to resolve them.

We would like you to consider two potential top jobs in headquarters depending on which way George and I—with your assistance—decide to organize our aeronautics work. If we leave aeronautics in OART, we would propose you consider the post of the Deputy Associate Administrator for Aeronautics in OART. If we decide to create a new Office of Aeronautics, we propose you consider the post of Deputy Associate Administrator in a new program office headed by a nationally known but not yet identified Associate Administrator for Aeronautics.

If this sounds reasonable to you, please give Betty Covert a call to set a date when you and George Low and I could get together to discuss the problem of aeronautics organization within NASA and the role which you would consider for yourself in this very important activity.

Sincerely,

Tom Paine
T. O. Paine
Administrator
NASA

Paine added in his handwriting: "Other individuals involved have not been told of this, so we would like you to consider this confidential."

Armstrong took a few weeks to consider the NASA administrator's request, but ultimately he felt he had little choice other than to accept the position Tom Paine was offering him, once it was clear that he would not be able to stay active as an astronaut and be assigned to another space mission. As Neil related in an interview for First Man, "I never asked the question about returning to spaceflight, but I began to believe that I wouldn't have another chance, although that never was explicitly stated."[23] The fact that his good friend George Low, since December 1969 the deputy administrator of NASA under Tom Paine, said he would like Neil to consider going back to aeronautics and take a deputy associate administrator job in Washington was the clincher. Though Neil was not convinced a job at NASA Headquarters would be right for him, he accepted the post as deputy associate administrator for aeronautics within the Office of Advanced Research and Technology (OART).

It must be emphasized that top NASA leadership was not simply putting its most famous astronaut and global icon on ice or protecting the First Man on the Moon from the dangers of another lunar mission. Aeronautics—the first "A" in NASA—was in fact instrumental to the agency, and to the country, in the early 1970s. Many new aeronautical technologies were coming down the pike, including digital fly-by-wire (DFBW), for which Neil was a strong advocate. Moreover, the fate of NASA's foremost aeronautical endeavor of the 1960s—the technologies needed for an effective commercial supersonic transport—was still hanging by a string in late 1970; in fact, a few months after Neil became the deputy associate administrator for aeronautics—the national SST program would fall victim to a spider's web of politics when, on March 24, 1971, in one of the most dramatic roll calls in modern U.S. Senate history, fifty-one senators voted to deny further funding. In August 1971, Neil resigned his NASA position. As he related in interviews for First Man: "I had always told people it was my intention to go back to the university. That was not a new thought for me. I didn't want to leave NASA precipitously, though it was never my intention to be in that bureaucracy job that long."[24] He joined the aerospace engineering faculty at the University *of Cincinnati for fall semester 1971. NASA seems to have offered little resistance to Armstrong's departure.*

"I AM SENDING YOU AN EXACT REPLICA"

July 20, 1970

Dear Neil:

One year ago today you and Buzz Aldrin emplaced on the lunar surface a small silicon disc bearing goodwill messages from the leaders of more than seventy nations around the world. On this anniversary of man's first landing on the moon, I am sending you an exact replica and an enlarged framed reproduction of that disc. Besides the goodwill messages, the disc contains excerpts from statements on America's goals in space by each of the four Presidents that have led the U.S. space program, and lists Congressional and NASA officials with important space program responsibilities.

As you know, the printing on the half-dollar-sized silicon disc was etched by a special microminiature electronic process developed to produce high reliability circuits for space flight. Each message was reduced to 200 times to a size smaller than the head of a pin and appears on the disc as a barely visible dot. Through a high-powered microscope the printing is sharp and clear, however. The messages and statements on the enlarged reproduction, where the messages are reduced only 10 times, can be read with the help of a small magnifying glass.

At the top of the disc is the inscription: "Goodwill messages from around the world brought to the moon by the astronauts of Apollo 11." Around the rim is the statement: "From planet earth—July 1969." The goodwill messages are in the center of the disc and the Presidential statements and the lists of Congressional and NASA officials are at the bottom.

I hope that this replica will give you pleasure as a reminder of the worldwide outpouring of goodwill to the United States engendered by the accomplishment in 1969 of America's first major goal in space. It is a personal pleasure for me to present it to you. I am also sending replicas to Buzz Aldrin and Mike Collins.

Sincerely yours,

Tom Paine
T. O. Paine
Administrator
NASA

"I ALMOST MADE A 'GIANT LEAP' OF MY OWN"

October 5, 1970

Dear Neil,

I stopped by CIN [Cincinnati] last Sat. for the presentation at the Airshow. Went very well, and glad to see your brother again.

    I've enclosed a program because I thought you would get a chuckle out of the picture inside the back cover. Where they got hold of that Choco Hilton picture I'll never know, but I'd like to have one of those if NASA has that one on file. Would appreciate it very much if you could have someone check with Julian's people. I still remember when that first Choco face peered around the corner. I almost made a "giant leap" of my own.

Best regards,

John
John Glenn
Columbus, Ohio

In this letter to Armstrong, John Glenn is making reference to the jungle survival training that he and Neil did together in June 1963. The training was organized for the astronauts by the USAF Tropical Survival School at Albrook AFB in the Panama Canal Zone. "Choco" was a member of the local Choco tribe who surprised them with a visit while they were encamped. After they had built their two-man lean-to of wood and jungle vines, Neil used a charred stick to write the name "Choco Hilton" on it.

    An American hero and one of the most popular astronauts of all time, John Glenn was a member of the original Mercury Seven (Original Seven, NASA Group 1, 1959). On February 20, 1962, he flew three orbits in Mercury-Atlas 6,

FELLOW ASTRONAUTS AND THE WORLD OF FLIGHT

becoming the first U.S. astronaut to orbit the Earth. He resigned from the astronaut corps in January 1964. Ten years later he won election as U.S. Senator from the state of Ohio, ultimately serving four consecutive terms. In October 1998, at age seventy-seven, Glenn returned to space as a member of the STS-95 crew, during which a series of medical experiments were conducted on Glenn studying the aging process and the problems of bone and muscle loss, balance disorders, and sleep disturbances.

Neil and John stayed good friends right up to the time of Neil's death in August 2012. John lived for four more years, dying on December 8, 2016.

"Watch me maneuver the mahogany desk"

October 13, 1970

Mr. John Glenn
Columbus, Ohio

Dear John:

Glad to hear the air show and attendant ceremonies in Cincinnati went well and that you were able to keep my brother under control.

I appreciate your sending the program and was pleased to see that picture. I don't have a copy of it myself but will try to trace it through MSC [Manned Spacecraft Center] and send you a print.

Stop by and watch me maneuver the mahogany desk when you come to Washington the next time.

Best wishes.

Sincerely,

Neil A. Armstrong
Deputy Associate Administrator for Aeronautics
Office of Advanced Research and Technology
NASA Headquarters

"WE WISH TO CORRECT AN INJUSTICE"

November 13, 1970

Mr. Arthur Ochs Sulzberger
President and Publisher
The New York Times
New York, New York

Dear Sir:

We wish to correct an injustice done to Neil A. Armstrong, Associate Director for Aeronautics, National Aeronautics and Space Administration, in a *Times* story of November 11, 1970.

The story by your correspondent, Christopher Lydon, with an accompanying picture caption of Mr. Armstrong, reported that "Mr. Armstrong was seeking to rally support for American SST program, now stalled in Congress." This is in error.

Mr. Armstrong appeared at a luncheon as the guest of the Aviation/Space Writers Association and the Aero Club of Washington to report on the technical aspects of his recent visit to the Tupolev Design Bureau in the U.S.S.R. Robert Hotz, Editor-in-Chief of Aviation Week & Space Technology, provided an historical analysis of Soviet civil aviation.

Mr. Armstrong's recorded remarks will confirm clearly that he addressed himself only to the technology of the Soviet SST and its significance.

Mr. Armstrong was invited to speak as a skilled aerodynamicist in the spirit of *whatever* he had to say about the state of the Soviet civil aviation effort would be important news. It is a distortion to report to the Times readers that Mr. Armstrong "was seeking to rally support for American SST program . . ."

The Aviation/Space Writers Association has a mission to report SST developments. With the trade of journalism coming under increasing attack by persons who are politically motivated, it is incumbent upon us as journalists to police our professional ethics.

Sincerely,

Jerry Hannifin
Time Magazine

Vern Haugland
The Associated Press

David Brown
Aviation Week & Space Technology

J. S. Butz
Nation's Business

"Grossly unfair representation"

December 4, 1970

Messrs. Jerry Hannifin, Vern Haugland, David Brown, and J. S. Butz
Members, Aviation/Space Writers Association
Washington, D.C.

Gentlemen:

I was very pleased to receive a copy of the letter that you jointly forwarded to the New York Times. I had seen the Times story and felt that it was a grossly unfair representation of my remarks and their intent.

I certainly appreciate your stand in bringing your views to the eyes of the publisher and congratulate each of you for your defense of quality journalism. Thank you very much.

Sincerely,

Neil A. Armstrong
Deputy Associate Administrator for Aeronautics
Office of Advanced Research and Technology
NASA Headquarters
Washington, D.C.

The development of an American Supersonic Transport (SST) was NASA's foremost aeronautical endeavor of the 1960s. The program fell victim to politics, however, when on March 24, 1971, in one of the most dramatic roll calls in modern U.S. Senate history, fifty-one senators voted to deny further funding for the American SST program. Although Armstrong's office had not had any responsibility for the SST program, Neil followed it closely and was in favor of continuing the program. He thought the prototype aircraft Boeing was building would be a good research machine that would benefit NASA research. Neil knew William M. Magruder, the national director of the SST program, first for the FAA and then for the Department of Transportation, very well; Neil had a lot of responsibilities in that area as NASA's deputy associate administrator for aeronautics. Former NACA/NASA/North American test pilot Scott Crossfield was the SST guy for Eastern Airlines at the time, and Neil knew him as well. Different congressmen and senators asked for Neil's opinion on SST matters. Wisconsin senator Gaylord Nelson, who had introduced legislation to prohibit the operation of any civil supersonic aircraft within the territorial jurisdiction of the United States until—and unless—sonic boom and stratospheric pollution from such aircraft could be reduced to zero, asked Neil a specific question about the subject (Armstrong was in favor of continuing the SST program) and Neil's answer wasn't to his liking. Nonetheless, Nelson went right over to the floor of the Senate and immediately quoted Neil as saying just the opposite of what he had said. Neil was learning the ways of Washington.

The article in the New York Times *on November 13, 1970, came nearly four months before the Senate's killing of the SST program on March 24, 1971.*

"IF YOU WOULD BE INTERESTED IN HAVING A PIECE AS A MEMENTO"

December 17, 1970

Dear Neil:

A friend of mine at MSFC [Marshall Space Flight Center] by the name of Bob Schwinghamer was in the office the other day and he had some parts of your booster that had fallen on a German freighter just after you went into orbit. These are the only parts recovered from any Saturn V and he wondered if you would be interested in having a piece as a memento. I told him you would so here they are. I am also enclosing a copy of his letter.

   Hope to see you soon.

Warmest regards,

Alan
Alan L. Bean
Captain, USN
NASA Astronaut

FELLOW ASTRONAUTS AND THE WORLD OF FLIGHT

Alan Bean became an astronaut in 1963 as part of Group 3. He served as the lunar module pilot for the lunar landing mission of Apollo 12 in November 1969 and was the commander of the fifty-nine-day Skylab 3 mission, which lasted from July 28 to September 25, 1973. In the two years before the launch of Apollo 11, Bean was Neil's office mate in the Astronaut Office at the Manned Spacecraft Center in Houston.

"MIGHT MAKE ME START DRINKING"

January 15, 1971

Mr. H. Russell Hair
Assistant for Apollo, Code DA
Medical Research and Operations Directorate
NASA Manned Spacecraft Center
Houston, Texas

Dear Russ:

Please forgive my late reply to your kind letter. I managed to get a couple of weeks off over the holidays and am still way behind in the correspondence department.

Janet reported on your gathering prior to her departure. I'm sorry I missed it.

I don't become generally involved in the selection of which speaking engagements to accept and which to decline. As you can imagine, that might make me start drinking. If your friend wishes to send an invitation, however, I would be most pleased to forward it to the appropriate office for prompt action. I really don't want to sound encouraging, however, as I believe they will not accept any additional commitments in this time period.

I hope to get down to Houston in the near future; and, if so, I'll drop in on Marcella.

All the best.

Sincerely,

Neil A. Armstrong
Deputy Associate Administrator for Aeronautics

Office of Advanced Research and Technology
NASA Headquarters
Washington, D.C.

"THE UPGRADING OF JAMES A. LOVELL"

May 12, 1971

Membership Committee
The Society of Experimental Test Pilots
Lancaster, California

It is my pleasure to recommend the approval of the upgrading of James A. Lovell to Associate Fellow in The Society of Experimental Test Pilots.

    Jim, as you know, is the most experienced space crewmen in both number of flights and space flight hours in the world. More than that, his participation in the planning and testing related to his own flights and others have been a significant contribution to the over-all success of the U.S. Manned Space Flight Program.

    In my view, there is no question that the Society should confer on him every honor that his accomplishments deserve.

Neil A. Armstrong
Deputy Associate Administrator for Aeronautics
Office of Advanced Research and Technology
NASA Headquarters
Washington, D.C.

Neil's letter of support for Jim Lovell came thirteen months after the ill-fated Apollo 13 mission, which Lovell commanded and managed to return to Earth safely following a critical spacecraft failure on the way to the Moon. Lovell had also served as the command module pilot for Apollo 8, which circumnavigated the Moon in December 1968, and flown on Gemini VII in December 1965 and Gemini XII in November 1966. He was also Neil's backup commander for Apollo 11. By the time Neil wrote this recommendation in May 1971, Lovell had become the first person to fly in space four times, and the only one to have flown to the Moon twice without making a landing.

FELLOW ASTRONAUTS AND THE WORLD OF FLIGHT

"SORRY TO LEARN THAT YOU WILL BE LEAVING NASA"

September 14, 1971

Dear Neil:

I was surprised and sorry to learn that you will be leaving NASA; however my regret is quite strongly tempered by the knowledge that you will be active in education. One of the very difficult aspects of the decision I made two and a half years ago to join NASA was that I was very reluctant to completely leave university work. Fortunately, I was able to persuade the University of California, Davis campus, to keep me on as a part-time instructor. I will give one regular course every quarter and have lots of fun doing it. I can therefore say, at least as a part-time contributor, "Welcome to the club."

Seriously, I think you will find your new work very challenging and important. Students, because they are not inhibited by more "mature" considerations, tend to ask the difficult and important questions. Every time I teach a class, I feel special responsibility to try and anticipate and then to answer the sharp questions that I know are forthcoming.

If you have an opportunity I would very much like to sit down with you the next time I am in Washington and talk a little about the problems now facing higher education.

With best personal regards,

Sincerely yours,

Hans Mark
Lecturer
University of California, Davis

P.S. Enclosed is a brochure describing the activities of the Department of Applied Science—you might be interested in looking it over.

Hans Mark (born in Mannheim, Germany, in June 1929) served as the director of NASA's Ames Research Center in Mountain View, California, from 1969 to 1971. In 1977 Dr. Mark became director of the National Reconnaissance Office (NRO) and in 1979 went on to serve for two years as secretary of the air force. In July 1981 President Reagan appointed Dr. Mark as deputy administrator of NASA, a post he held until September 1984, at which time he became the chancellor of the University of Texas. In 1992 he left administrative work to teach as senior professor of aerospace engineering on the Austin campus. In July 1998 President Clinton named him director of defense research and engineering in the Pentagon. He returned to teaching at the University of Texas in 2001, retiring in 2014.

"I HAVE A BIT OF A DILEMMA"

January 6, 1972

Dear Neil:

I hope that all is well in the University of Cincinnati and that the aero students aren't falling behind too rapidly!

I have a bit of a dilemma. The guy wiring the attached [letter] works with me on an advisory group formed between the National Park Service and the citizen (users). He is an eager young environmentalist with slight touches of "Sierra Club fever" who knows we have been together in Washington. I certainly did not want to give him your address without your approval, so I was a bit evasive when he requested it at our last meeting. He has taken the tack of routing a letter to you thru me. Rather than pass him your address as he has requested in his P.S., I'm sending this on to you for your choice of disposal. If you choose not to respond, and I wouldn't blame you for doing it, I can always cover by saying it was lost in the shuffle of your move.

My best to the family and please feel free to stay with us during your visits to D.C.

Sincerely,

Bill
William A. Anders
Executive Secretary
Executive Office of the President
National Aeronautics and Space Council
Washington, D.C.

William A. Anders (b. 1933) was part of NASA's third group of astronauts, selected in 1962. After serving as the backup pilot for Gemini XI, he flew, along with Frank Borman and Jim Lovell, in December 1968 as lunar module pilot for Apollo 8, a historic circumlunar flight that was the first mission in which humans traveled beyond low Earth orbit. Bill also served as the backup command module pilot for Apollo 11. Though he maintained his astronaut status, he took an assignment in 1971 with the National Aeronautics and Space Council, a body that was responsible to the president for developing policy options concerning research, development, operations, and planning of aeronautical and space systems. In 1973 he was named to the five-member Atomic Energy Commission, during which time he served as the American chair of a joint U.S./USSR technology exchange program for fission and fusion power. In 1975 Anders was named by President Ford as the first chair of the newly established Nuclear Regulatory Commission, which was responsible for nuclear safety and environmental compatibility, a responsibility he held for only a short time before becoming ambassador to Norway. He ended his career in the federal government in 1977 and began work in the private sector, with General Electric, Textron, and General Dynamics. He retired in 1993 but consulted part-time with the U.S. Office of Science and Technology Policy and was a member of the Defense Science Board and the NASA Advisory Council. He retired from the air force as a major general. In 2005 he established the William A. Anders Foundation, a philanthropic organization dedicated to supporting educational and environmental issues.

"I TOOK THE LIBERTY OF TELLING THEM THAT I THOUGHT IT WAS NOT POSSIBLE"

January 24, 1972

Dear Neil:

On Saturday, February 26, the local school district will hold the formal dedication of the Joe Walker Junior High School with a small ceremony. For obvious reasons, we are assisting wherever possible.

They of course wanted you to participate in the ceremonies, but knowing of your schedule, I took the liberty of telling them that I thought it was not possible.

However, I did want to let you know about it. I hope I was not too presumptuous in declining for you, and if I was in error, I am sure I can easily rectify it.

If you wish to send a telegram or message of some kind to the school I will be happy to see that they get it, or if there is anything else you would like to do, we will be happy to help.

I'll bet you're glad to get away from Washington. I don't see how you took it as long as you did. As far as I can tell, my uncle Roy is still trying to run the world.

Best Regards,

Ralph B. Jackson
Public Affairs Officer
NASA Flight Research Center
Edwards, California

Joseph A. Walker (1921–1966) had been one of Neil's closest friends, as they spent five years together as NACA/NASA flight research test pilots at Edwards Air Force Base in California. Walker had been killed in June 1966 in a freak midair collision over the Mojave Desert. It happened when Walker's F-104N Starfighter inexplicably flew too close to a plane with which he was flying in formation—the XB-70A Valkyrie, a $500 million experimental bomber that North American Aviation had designed for Mach 3–plus speeds—and became caught in the mammoth plane's extraordinarily powerful wingtip vortex. Walker died instantly. One of the Valkyrie pilots, air force major Carl S. Cross, died in the wreckage of the bomber. The other XB-70A pilot, Al White, a test pilot for North American, survived via the plane's ejection capsule, but not without some serious injuries. Magnifying the tragedy was that the deaths came during what amounted to a publicity shoot for General Electric.

"I would like to be included"

February 1, 1972

TELEGRAM

Mr. Ralph B. Jackson
Public Affairs Officer
Flight Research Center—NASA
Edwards, California

FELLOW ASTRONAUTS AND THE WORLD OF FLIGHT

I WOULD LIKE TO BE INCLUDED AMONG THOSE WHO WOULD WISH TO BE PRESENT PERSONALLY ON THE OCCASION OF THE DEDICATION OF JOE WALKER HIGH SCHOOL. JOE LEFT A PERSONAL AND PROFESSIONAL LEGACY TO THIS COMMUNITY OF WHICH THOSE OF US WHO HAD THE PRIVILEGE OF KNOWING HIM AS A FRIEND CAN JUSTLY BE PROUD. I COMMEND THE COMMUNITY FOR THE APPROPRIATENESS OF THIS CEREMONY AND SEND MY SINCERE BEST WISHES FOR THE FUTURE SUCCESS OF JOE WALKER JUNIOR HIGH.

NEIL A. ARMSTRONG

"LONG TIME SINCE OUR LAST VISIT WITH PAT"

August 20, 1973

Dear Jan and Neil:

I have been browsing through my space memorabilia and reminiscing about the past. It seems a long time since our last visit with Pat and my chores of raking and cleaning up the yard. Now most of our best friends have left NASA and Pat is happily married in Houston. She and Lloyd came through St. Petersburg last year and we were very pleased with their new life together. They have been in Europe and Africa for the last eight weeks and are just getting home. Lloyd has a ranch in South Africa and they took Bonnie with them on a rather fancy safari.

I was thumbing through my first day space covers and found that I do not have your autograph on my two most important ones—your famous flight to the moon, Apollo-11, and the cover commemorating your taking the Apollo-1 arm patch to the surface of the moon. These are part of my own personal collection which I shall someday probably pass on to Ed III or Bonnie. I'd appreciate your signing each one and then I'll send them along to Mike and Buzz.

I am on the Board of Directors of the National League of Families of POWs and MIAs. The Board meets in Washington this weekend and Mary and I are flying in to try to dig out some information about our

Jim and others who were reported down over Laos. I feel strongly that the Pathet Lao is holding several hundred of our MIAs in secret camps and perhaps in China. It's a delicate situation!

Does the locals and subjects of the enclosed picture look familiar? I remember the convenient hole in the fence that made quick access possible.

We hope you are both enjoying your activities in Cincinnati. I follow the fortunes of the Reds but that is as close as we get nowadays.

Cordially,

Mary and Ed White

Mary Rosina White (née Haller; 1900–1983) and Edward H. White Sr. (1901–1978) were the parents of Edward Higgins White II, the NASA astronaut (Group 2) who was the first American to make a spacewalk (Gemini IV, June 1965) and who died, along with Apollo 1 crewmates Gus Grissom and Roger Chaffee, in the Apollo fire of January 1967. In Mary Rosina's letter to Neil, she mentions her other son, James B. White, an air force pilot who was shot down over Laos during the Vietnam War. Captain White's plane, following a mission over Laos in November 1969, failed to return to base and was classing as MIA. Because Laos was officially neutral, and because the U.S. continued to state it was not at war with Laos (although the U.S. regularly bombed North Vietnamese traffic along the border and conducted assaults against communist strongholds in Laos at the behest of the anti-communist Laotian government), the U.S. did not negotiate for Americans lost in Laos. At war's end, no American held in Laos was released. For years, the White family believed that Jim was still alive in captivity in Southeast Asia along with hundreds of other Americans.

In July 2017, the disappearance of James B. White was finally accounted for. On November 24, 1969, Capt. White, a member of the 357th Tactical Fighter Squadron, was piloting an F-105D aircraft in a flight attacking enemy troops. During the mission, weather conditions deteriorated and contact with White was lost. On November 28, an Air America helicopter sighted wreckage thought to be White's aircraft. A Laotian ground team searched the area and found small pieces of wreckage, but no remains were recovered.

FELLOW ASTRONAUTS AND THE WORLD OF FLIGHT

"Happy to break my rule in this one instance"

October 15, 1973

Maj. Gen. and Mrs. Edward H. White, USAF (Ret.)
St. Petersburg, Florida

Dear General and Mrs. White:

It was a pleasure to find your letter awaiting my return to the campus after a pleasant summer on our farm. We are on the quarter system and don't begin classes till October.

I have not signed first day covers for some years because of their commercial negotiability. I am, however, happy to break my rule in this one instance, knowing the care that you will give to them.

Thanks for the picture from Woodland Drive. We remember our days there warmly. We have not seen Pat in a long time but hope she is happy in her new life.

We are very much enjoying our life on the farm. It is an active working farm and the boys really enjoy having the running room.

I hope you'll be able to stop and see us sometime. Our farm, Rivendell, is 3 miles Northeast of Lebanon Ohio on Route 123.

Sincerely,

Neil A. Armstrong
Professor of Aerospace Engineering
University of Cincinnati

The Whites (Ed White III, wife Pat, daughter Bonnie, and son Eddie III) and the Armstrongs had arrived together in Houston in the fall of 1962 as members of the New Nine. After living in rentals for over a year, the two families bought property together in the El Lago development, one of a handful of planned communities that had sprung up around the space center, scattered with ranch houses and crisscrossed with tidy streets. The families bought three contiguous lots and split the middle one in half so that they each had a lot and a half to build their house on. The two families naturally grew quite close. Separating the two backyards was a six-foot-tall wooden fence but, as the letters indicate, there was "a convenient hole in the fence that made quick access possible."

Because Janet Armstrong was Pat White's neighbor and friend, NASA requested that she be the one to prepare Pat for the news that her husband had just been killed in the Apollo fire. Reading between the lines in the above letters, one senses that Pat White had a very difficult time emotionally following her husband's tragic death, which is known to be true from other sources. Born in 1933, Patricia Finegan White Davis died in 1983 following a bout with cancer. The Lloyd she had married in the early 1970s was Lloyd Davis, president of Fisk Electric Company, based in Houston.

The name the Armstrongs chose for their farm in Lebanon, Ohio, Rivendell, was the name of the idyllic secluded valley of J. R. R. Tolkien's fictional Middle Earth in Lord of the Rings *and the abode ("the last homely house") of noble Elrond, who is half elf and half human. In the Rings trilogy Rivendell is the last place where elves live before leaving Middle Earth and returning to "the immortal lands" over the sea.*

"I SHALL ALWAYS BE GRATEFUL FOR THE OPPORTUNITY OF A LIFETIME"

February 25, 1974

Dear Neil,

I am sending you a copy of my book, "To Rule the Night." I want you to know first-hand what that space trip meant to me.

I also want to thank you for your help in getting me there. I shall always be grateful for the opportunity of a lifetime.

I am sending you a brochure on the High Flight Foundation. Reading it will give you an appreciation for what we are trying to do. We ask for your prayers and participation as you feel directed. I believe High Flight can be the most potent force to get our country back on the right path if you will lend your support.

Someday I hope we can all meet in a private setting where we can freely exchange our deepest feelings related to our space endeavors. Let us meet before it is all forgotten.

Space was a challenge to us all. I believe that we find out greatest challenge on the earth trying to communicate to others our ideas and aspirations. Let us work together to lift man's spirit.

With dreams of yesterday but hope for tomorrow,

Jim Irwin
President
High Flight Foundation
Colorado Springs, Colorado

James Benson Irwin (1930–1991) served as lunar module pilot for Apollo 15, the fourth human lunar landing and was the eighth person to walk on the Moon. After retiring as an air force colonel in 1972, Irwin founded the High Flight Foundation. He spent the remainder of his life as a self-declared "Goodwill Ambassador for the Prince of Peace" and spoke about how his experiences in space had made God more real to him.

"Contribute to a better tomorrow, each in our own way"

February 28, 1974

Mr. James Irwin, President
High Flight Foundation
Colorado Springs, Colorado

Dear Jim:

Thanks for your kind note and thoughtfulness in sending your book and the High Flight information.

I'm sorry I missed you when you visited Cincinnati on your book promotion tour. I don't travel much now, but I did happen to be away on that occasion. Anyway, thanks for calling.

We all try to contribute to a better tomorrow, each in our own way. I hope we all achieve some measure of success in that regard.

Janet joins me in sending you and Mary our very best. I hope our paths cross soon.

Sincerely,

Neil A. Armstrong
Professor of Aerospace Engineering

A RELUCTANT ICON: LETTERS TO NEIL ARMSTRONG

Armstrong's correspondence with former astronaut Jim Irwin about his High Flight Foundation provides some of the most interesting insights into Neil's position on religion and spirituality that exist. As readers will see, Neil wanted no part of Irwin's organization and did not share Jim's views on the Christian meaning of space exploration.

"Private meeting of those who have been on the Moon?"

April 21, 1977

Dear Neil,

I am writing this letter from Fitzsimons Army Medical Center. I have spent a lot of time in the hospital this year with heart surgery in January and another heart attack in March. I have had a lot of time to think. Since I might be the first to go, I want to propose an idea that has been on my mind for several years.

    Have you considered a private meeting of those who have been on the moon? As you know we comprise a very lucky, unique group of twelve individuals. I think it is very unfortunate that we have gone our own way with no apparent attempt to come together. I know very well that we are all different but we have shared a common experience. Eventually I would like to include all those who have flown to the moon but that must wait.

    If you agree on the desirability of such a meeting, we could use the facilities of the Viking Mountain Resort in North Carolina. This exquisite retreat is owned by Georgia Senator Eugene Holley and he offered the exclusive use for our get together. Air transportation would also be provided to the private resort.

    I would like to hear from you regarding the idea. If you agree that we should get together, then suggest a time or times when you could be away for a few days. Should wives be included? I would say no. If you would like information on the special quality of Viking Mountain, I will send that.

    When we get together, and I hope that we can, I would propose relaxation and some serious discussion. Again, I would emphasize there would be no one there but us.

My warm regards,

Jim Irwin
Chairman of the Board and Chief Executive Officer
Apollo 15 Astronaut
High Flight Foundation
Colorado Springs, Colorado

"Might be better to avoid any political connections"

May 11, 1977

Mr. James B. Irwin
Chairman of the Board
High Flight Foundation
Colorado Springs, Colorado

Dear Jim:

I was saddened to hear of your continued heart problems and do hope that the situation is improving and under control.

I have thought from time to time of a reunion: I'm certain all of us have. I do agree that it's a worthy idea, although I am not convinced that the CMP's [command module pilots], or for that matter, any of the Apollo crewmen should be excluded.

I do not know Senator Holley, nor am I suspicious, but I suggest that it might be better to avoid any political connections.

I'll think about it. Perhaps some additional thoughts will come to me.

I do hope your health continues to improve. All good wishes.

Sincerely,

Neil A. Armstrong
Professor of Aerospace Engineering & Applied Mechanics
University of Cincinnati

A RELUCTANT ICON: LETTERS TO NEIL ARMSTRONG

"A new appreciation for freedom"

August 8, 1978

Dear Neil,

I recently returned from a two-week speaking tour of the Soviet Union. Many of you have had the opportunity to visit Russia, but this was my first and perhaps my last visit. It was a very enlightening experience. First of all, I was surprised that my visa was approved. You will receive a trip report in the next issue of APOGEE. Needless to say, I came out with a new appreciation for freedom.

I found a great hunger in Russia for information. A frequently asked question concerned the spiritual lives of the other astronauts. I made no attempt to answer for the group, other than to say there are many dedicated Christians in the astronaut corps. I know there are many different views within our group.

I am writing this letter to each of you who has flown in space, requesting that you make a brief statement about your faith in God, the creation of the universe, or just your spiritual beliefs. There is a worldwide interest in your feelings on these matters.

I would also like your permission to share your statement.

I look forward to seeing you later this month in Houston during the briefing. Perhaps I can obtain the information requested from you at that time.

Warm regards,

Jim Irwin
Chairman of the Board and Chief Executive Officer
Apollo 15 Astronaut
High Flight Foundation
Colorado Springs, Colorado

APOGEE *was a bi-monthly newsletter of Jim Irwin's High Flight Foundation. It was edited by, among others, Ms. Mavis Sanders, who spent nearly forty years in Christian periodical and book publishing.*

The briefing in Houston mentioned in Irwin's letter was an event organized by Christopher C. Kraft, the director of Johnson Space Center. Kraft's idea was to provide technical briefings to NASA's former astronauts and get their input.

FELLOW ASTRONAUTS AND THE WORLD OF FLIGHT

NASA administrator Robert Frosch presented an overview at the August 1978 event and John Yardley, associate administrator for the shuttle program, discussed the new space transportation system (space shuttle). Glynn Lunney, Tom Stafford, Deke Slayton, and John Young also briefed the former astronauts on the design and operational features of NASA's new orbiter. Twenty-two former astronauts attended. In the group picture Neil Armstrong chose to stand in the very back row, on the end. Mike Collins and Buzz Aldrin stood in the front.

"Would not choose to present my personal beliefs in any public manner"

November 21, 1978

Mr. James B. Irwin
President
High Flight Foundation
Colorado Springs, Colorado

Dear Jim:

Receiving a copy of your latest flyer concerning your visit to the USSR reminded me that we had not had the opportunity to visit in Houston, primarily due to my abbreviated schedule. In particular, we did not have the opportunity to discuss your earlier inquiry regarding making a statement available concerning individual spiritual beliefs.

As for myself, I would not choose to present my personal beliefs in any public manner. I admit to some difficulty in determining the appropriateness of using the public exposure garnered as a result of NASA activities to influence others in commercial, political, or spiritual matters.

I have no desire to avoid reality. I do know that we have some significant power to influence, and have the obligation to use that power carefully.

In any case, I hope our paths cross more often in the future.
All good wishes.

Sincerely,

Neil A. Armstrong
Professor of Aerospace Engineering & Applied Mechanics
University of Cincinnati

A RELUCTANT ICON: LETTERS TO NEIL ARMSTRONG

"You have a great power of influence"

November 18, 1979

Dear Neil,

Just last week my Father passed away. As I prepared my remarks for his memorial service, I re-evaluated my own life. I wonder whether there is more that I can do to preserve the values we treasure as important.

Perhaps you have had similar thoughts? I still believe you have a great power of influence, because you were once in space. You are adding to that influence by your present endeavors. Some people just want to view us a little differently.

I would like you to consider the possibilities of an astronaut forum as your voice to the American people. This effort would not be connected to High Flight Foundation in any way. I realize that some of you would not care to publicly comment on certain issues, and in some cases could not because of your business position.

A recent issue that I might use, for example, is the Taiwan action. Frankly, I am embarrassed that the United States would desert this friendly country to court Communist China. Is loyalty to a friend just a sentimentality of the poet? I believe our President acted unwisely. Can our other allies now expect us to give them the same shoddy and shameful treatment if we forsake loyalty for business interests? I guess I feel very strongly because I visited Taiwan in 1972, and my brother's last Air Force assignment was there.

What can we do for America? We can do a great deal by speaking out on vital issues that will affect our future and the future of America. Would you be willing to help by participating in such an effort as the Astronaut Forum?

With great expectation,

James B. Irwin
President
High Flight Foundation
Colorado Springs, Colorado

FELLOW ASTRONAUTS AND THE WORLD OF FLIGHT

"Not persuaded that the general good would be well served"

March 22, 1980

Mr. James B. Irwin
President
High Flight Foundation
Colorado Springs, Colorado

Dear Jim:

Thank you for your thoughtful letter inviting comment on the possibility of an "Astronaut Forum."

Our colleagues are intelligent and considerate observers of the world scene. They share a perspective that is unique. It does not necessarily follow, however, that their collective opinion on important matters would be superior to that of other groups with carefully reasoned opinions, or, for that matter, the public at large.

I am not persuaded that the general good would be well served by public pronouncements of collective astronaut opinion. Individual views, on the other hand, should and will be accepted with the "grain of salt" that is probably appropriate.

It's just one opinion, Jim. I'm certain we all won't agree on this one either.

Janet and I send sincere condolence on your father's passing, but best wishes for the happiness of all the Irwin family.

Neil A. Armstrong
Professor of Aerospace Engineering & Applied Mechanics
University of Cincinnati

"Encourage each of you to write your impressions"

August 3, 1984

Dear Neil,

I am sorry I was unable to join you for the special reception at the White House on 20 July 1984. I had obligated myself to speak on that day for the French Aviation and Space enthusiasts gathered at Meribel in the

French Alps. Not a bad assignment at all. It was a wonderful week for Mary and me.

While there the subject of reflections on the moon visit was probed by news media. The prime mover for the festival at Meribel, "Semaine Sur La Lune," was journalist and writer Bernard Chabbert. He asked if we had written our personal reflections of the moon. I know that many have, but I am prompted to encourage each of you to write your impressions not necessarily to be published but just to be preserved for the future.

The astronauts from Western Europe were also there and they are very interested in encouraging dialogue with others who have or will fly in space. Perhaps they are the ones who could take leadership to bring it all together.

I regarded the visit to the French Alps as opportunity to condition for Mt. Ararat but there were more important results.

I wish you and your family the very best for life.

With great appreciation,

Jim
James B. Irwin
President
High Flight Foundation
Colorado Springs, Colorado

Starting in 1973, Jim Irwin led several expeditions to Mount Ararat in Turkey, in search of the remains of Noah's Ark. In his 1983 book More Than Earthlings: An Astronaut's Thoughts for Christ-Centered Living, *he expressed his view that the Genesis creation story was literal history. Jim died on August 8, 1991, after his fourth heart attack, the first of which had occurred less than two years after Apollo 15.*

"If anyone can put it together you can"

October 4, 1991

Dear Neil,

Thank you for your kind letter of sympathy. Actually your letter cleared up a puzzling situation as the enclosed letter will reveal. Looking back on it now with a clearer head and with my emotions nearly back to normal, I find the "impostured" letter quite annoying. Mostly from the standpoint that someone would play on the emotions of a grieving widow and her family and then sign your name to it. I admit, because I don't know you very well, I nearly fell for it. Do what you want with the letter.

Since you couldn't attend Jim's service, I will be happy to send you a video of the service under a separate cover. It is the most upbeat and joyful funeral I've ever attended. The power of God was so strong in the church I couldn't shed a tear. It's so difficult to try and explain, you'll have to see and hear it for yourself.

We shall carry on with the mission and goal of the High Flight Foundation. The board has elected me to fill the president's position. The responsibility nearly overwhelms me, but I know God is faithful and will stand at my side.

One of the last comments Jim made before he lost consciousness was the astronaut reunion he was planning. He found a lovely retreat facility for it and wanted so very much to have each one attend who had been moon-walkers. His desire was to document the spiritual overtones that each man felt concerning their flight. He seemed to feel an urgency in the timing which turned out he was right.

Neil, I do not know if you are interested in carrying out Jim's dream but if anyone can put it together you can. I approached Al Bean on the subject and he is willing to attend but felt he was unqualified. He was also concerned about the possibility of hurt feelings over not inviting all of the astronauts who had flown. This could be a problem and you would know better than I.

Take your time and think about it. If you should decide it's a good idea and want to pursue it I'll be happy to help you make the arrangements or do anything I can to assist.

Take care of your heart, there's only one you.

Mary Irwin
High Flight Foundation
Colorado Springs, Colorado

Neil had his own heart attack the same year Jim Irwin died from one, in February 1991. It happened on the ski slopes in Aspen, following his separation from Janet (they eventually divorced in 1994), on a vacation with his good friends from Upper Sandusky, Ohio, Kotcho and Doris Solacoff, along with Neil's brother, Dean, himself recently divorced. The ski patrol brought him down from the mountain in a rescue toboggan and from there Neil was taken to the doctor on duty at the lodge infirmary, who confirmed a heart attack and administered atropine through an IV line to stabilize his cardiac arrhythmia. An ambulance transported Neil to Aspen Valley Hospital, where he was placed in the intensive care unit. There Armstrong experienced repeated episodes of bradycardia, or abnormal slowing down of the heart, but his heart rate soon stabilized enough for a transfer to Denver, which did not occur for three days because a blizzard had set in. Kotcho Solacoff, himself a physician, arranged for a transport by medivac from Aspen to a hospital in Cincinnati. There a team of heart specialists carried out a catheterization that linked the attack to a tiny aberrant blood vessel. The rest of Neil's coronary arteries were clear of blockages of any kind; his heart tissue sustained only the slightest amount of permanent damage. Released the next day with no major restrictions, Armstrong took the heart specialist at his word and flew to a business meeting. Six months later, he passed his flight physical and was put back on full flight status.

Twenty-one years later, in August 2012, Neil died from complications following quadruple heart bypass surgery.

With her letter to Neil, Mary Irwin attached the original of the fraudulent letter sent to her under the name Neil Armstrong. It is a crazy letter and certainly does not merit publication in this collection.

"HAPPY TO KNOW THAT YOU AND JAN FINALLY FOUND SOME DEGREE OF TRANQUILITY"

September 13, 1974

Dear Neil,

I've been meaning since July to drop you a note of congratulations on the occasion of the fifth anniversary of the Eleven Mission. If things had worked out, I'd have made it back east either for the Cape ceremony or at Mike's place [Mike Collins] . . . though I dread going back to D.C. even for a visit. Unlike Bill Anders . . . I had my fill of that town when we were at the Space Council.

I thought of you recently when Charles Lindbergh passed away. I greatly appreciated even the few moments I shared chatting with him and I know that high on your list of achievements was the friendship that developed between you two.

It's amazing to me how many of the old gang pass through this cow town at one time or another . . . . from Lew Hartzel, Cernan . . . to Irwin and Lovell. Phoenix is great and I have no regrets about leaving the space program in '71.

Diane and I are happy to know that you and Jan finally found some degree of tranquility (no pun intended) . . . That Washington banquet circuit was above and beyond.

I've seen Buzz once or twice in California . . . . I just hope he finds the right road for him soon. Mike's book is really well written and I hope he does well with it.

In closing, it's still somewhat sad to me to see the slowdown of the program. I wonder if you remember a night or two just prior to Eleven's liftoff, when Bill Anders and I had dinner with you three in the quarters. We were talking about the budgetary slowdown we foresaw on the Hill and I gave you a small seashell from Cocoa Beach with a joking reference that if you'd throw it in with the first sack of rocks . . . we'd probably still be making landings!

We wish you health, happiness and peace . . . . .

Sincerely,

Chuck
Charles D. Friedlander
President
Western Ranchlands, Inc.
Scottsdale, Arizona

Charles D. Friedlander (b. 1928) directed the astronaut support office at Kennedy Space Center from 1963 to 1967, before moving on to serve through the first Moon landing as the space technical consultant for CBS News. From 1969 to 1972 Friedlander was executive assistant for the National Space Council. He earned a BS degree in engineering from the U.S. Military Academy in 1950.

"LET ME HEAR WHAT YOU HAVE BEEN DOING"

March 10, 1975

Dear Neil:

Happy 1975. I hope it is both a personally and professionally rewarding year for you. One of my resolutions this year is to stay in closer touch with old friends and acquaintances.

1974 was filled to overflowing with three months at Harvard Business School, writing a book, building a home, and changing my business affiliation from real estate to the energy industry. This last pursuit led me to join HydroTech International in September as General Manager of one of their subsidiaries. HydroTech is a five year old company addressing itself to the challenge of finding more efficient and cost effective solutions to expensive problems for the offshore pipeline industry. HydroTech Development Company is responsible for the initial hardware development that will provide those solutions. It is an exciting pleasure to be back with a technology oriented business.

Let me hear what you have been doing, and perhaps our paths will cross in 1975. If I can do anything for you in Houston, please let me know.

Sincerely,

Walt
Walter Cunningham

P.S. Best to Jan.

Walter Cunningham (b. 1932) was the lunar module pilot on the Apollo 7 mission in 1968. He was NASA's third civilian astronaut, after Neil Armstrong and Elliot See, though he, like Neil, had earlier been a military fighter pilot. The book he refers to is The All-American Boys, *a memoir coauthored by Mickey Herskowitz, which was not published until 1977.*

FELLOW ASTRONAUTS AND THE WORLD OF FLIGHT

"Pleased to see you back in the technical world"

October 26, 1975

Walter Cunningham
President
HydroTech Development Co.
Division of HydroTech International, Inc.
Houston, Texas

Dear Walt:

If you can write in March, I can answer in October. At least this note won't get mixed up with the Christmas cards.

I was most pleased to see you back in the technical world. It seems like an exciting business and I hope you find it a rewarding business.

Universities remain the same: overloaded and out of money. It hardly seems possible that I'm in my fifth year here.

If your travels bring you this way, please stop by. Janet and I would be delighted to see you again.

All the best.

Sincerely,

Neil A. Armstrong
Professor of Aerospace Engineering
University of Cincinnati

Interestingly, after Neil resigned from the faculty at the University of Cincinnati, he too, like Walt Cunningham, got involved in the energy business. Neil entered into a business partnership with his brother, Dean, and their second cousin Richard Teichgraber, which is discussed in the introduction to the following chapter.

"WIDESPREAD PUBLIC APATHY ABOUT THE FUTURE OF OUR NATIONAL SPACE PROGRAM"

March 17, 1975

Dear Neil:

It is my privilege to invite you to join a group of prominent Americans who will serve as the Board of Governors of a newly formed National Space Association, a charitable, non-profit organization of which I have the honor of being the first President.

I am sure you are sharing my concern that the widespread public apathy about the future of our national space program is depriving us of many great benefits now available as a result of the billions of dollars spent on space R & D during the past fifteen years.

The group of individuals who banded together to form the National Space Association believe that there is a need to provide the public with a voice in the direction of the space program. Properly directed and supported, the program we plan to implement can help resolve some of the pressing energy, food, economy and balance of trade problems with which we are faced today. In addition, the National Space Association will play a key role in providing necessary information to the public concerning the many peripheral benefits derived from space endeavors.

The Board of Governors has been created as an integral part of the National Space Association to provide new ideas relative to our efforts, as well as to help generate more involvement in the association and the Nation's space program. Because of this need, and your historic role as the first man on the moon, we are asking you to participate as a member of our Board of Governors at its first annual meeting, held in conjunction with the Apollo/Soyuz liftoff at Cape Canaveral currently scheduled for July 15, 1975.

It is my fervent hope that you will consider this participation positively.

Sincerely,

Dr. Wernher von Braun
President
National Space Association

Dr. Wernher von Braun (1912–1977) was one of the fathers of rocket technology and space sciences in Germany and the United States. He came to the U.S. along with a number of German rocket scientists, engineers, and technicians following World War II as part of the U.S. Army's Operation Paperclip. After several years of constant work on rocket development for the army at White Sands, New Mexico, and the Redstone Arsenal in Huntsville, Alabama, von Braun became the first director of the new NASA Marshall Spaceflight Center, also in

FELLOW ASTRONAUTS AND THE WORLD OF FLIGHT

Huntsville. He served in that position through the first few Moon landings before becoming deputy associate administrator for planning at NASA Headquarters. Retiring from government service in May 1972, he moved to Fairchild Industries in Maryland as a vice president. He died of cancer on June 16, 1977.

The Apollo-Soyuz mission (known as the Apollo-Soyuz Test Project, or ASTP) did in fact take place on July 15, 1974. It involved the docking of an Apollo command/service module with the Soviet Soyuz 19 and was the first joint U.S.-Soviet space flight, a symbol of the policy of détente that the two superpowers were pursuing at the time. In the views of many, the mission marked the end of the Space Race that had begun in 1957 with the Sputnik launch. It was the last manned U.S. space mission until the first space shuttle flight in April 1981. It was also Mercury Seven astronaut Deke Slayton's only space flight as he was grounded until 1972 for heart-related issues.

"Must discipline myself"

May 7, 1975

Dr. Wernher von Braun
President
National Space Association
L'Enfant Plaza
Washington, D.C.

Dear Wernher:

I was most honored to receive your kind invitation to join the Board of Governors of the National Space Association. I'm certain that this new organization will get off to a rapid start under your able leadership.

Before I come to a decision on the matter, I should very much appreciate knowing some of the details regarding the National Space Association. I certainly agree with the broad goals as you have outlined them and look forward to understanding the mechanisms by which they might be achieved.

I was most encouraged by your own interest in this new organization, knowing how busy your schedule is. Still, like yourself, I must discipline myself to avoid "biting off more than I can chew."

It's not necessary to answer personally; perhaps some staff assistance would suffice for the present.

Best personal regards.

Sincerely,

Neil A. Armstrong
Professor of Aerospace Engineering

Neil decided not to accept the invitation to join the board of the National Space Association, which in 1975 became the National Space Institute and is now known as the National Space Society, an "independent, educational, grassroots, non-profit organization dedicated to the creation of a spacefaring civilization."[25] Its board of governors has included Buzz Aldrin, Tom Hanks, Norman Augustine, General Simon "Pete" Worden, and several other notable figures.

Armstrong met and talked with von Braun only a few times during his years as an astronaut, and there seems to have been only a couple of letters exchanged between them in the years following Neil's departure from NASA and von Braun's death from cancer in 1977.

In 1987, following the space shuttle Challenger *disaster, the NSI merged with the L5 Society based on the ideas of Princeton physicist and futurist Dr. Gerard K. O'Neill to create the National Space Society. Their merger proclamation read:*

> Whereas, the United States' space efforts have been most successful when they have reflected the dreams, values, and hopes of the American people;
>
> Whereas, recent difficulties in our national space program are largely due to a lack of public participation in the setting of national space goals and policy in previous years;
>
> Whereas, the National Commission on Space heard citizens across the country articulate their common vision of an open frontier in space for all humanity;
>
> Resolved, that:
>
> We, the former Presidents of the L5 Society and the National Space Institute, do announce and proclaim the merging of our organizations to become the National Space Society, and pledge to each other and to our members that we shall lead the American people in fully expressing their vision and more effectively participating in our nation's future in space.
>
> We further declare the Society's long-range goals to be: to create a spacefaring civilization which will establish communities beyond

the Earth; to promote the exploration and economic development of space; to advocate the opening of the space frontier; thereby enabling humanity to take its rightful place in the Solar System.

Ben Bova
National Space Institute
March 28, 1987

Gordon Woodcock
L5 Society
March 28, 1987

Today, the National Space Society has chapters in the United States and around the world. It publishes Ad Astra *magazine, an award-winning periodical chronicling the most important developments in space. NSS also organizes an annual International Space Development Conference.*

The National Commission on Space mentioned in the Bova-Woodcock merger proclamation of 1987 refers to a special commission of space experts that was created by President Ronald Reagan's executive order of October 12, 1984, to investigate and evaluate the future of the U.S. national space program. Heading that commission was Dr. Thomas O. Paine, a former NASA administrator. Armstrong served on the fifteen-person commission, along with Dr. Luis Alvarez, a winner of the Nobel Prize in Physics; Dr. Gerard K. O'Neill; Dr. Kathryn D. Sullivan, a space shuttle astronaut and the first American woman to walk in space; Dr. Jeane Kirkpatrick, former U.S. ambassador to the United Nations; and Brigadier General Charles E. "Chuck" Yeager, the air force test pilot who was the first to fly a plane beyond the sound barrier. The commission's report was entitled "Pioneering the Space Frontier." Published in February 1986, a month after the Challenger *accident, the report had little chance for impacting actual U.S. space policy in the aftermath of* Challenger, *as the cornerstone of the report was the assumption of shuttle-based cheap, safe, and reliable access to space. From that cornerstone, the Paine Commission report set as America's goal a manned landing on Mars in 2026, the cost of which it estimated at—overall, for establishing the necessary infrastructure in the 1995–2025 period—$700 billion.*

A RELUCTANT ICON: LETTERS TO NEIL ARMSTRONG

"IT LOOKS AS THOUGH THE SHUTTLE IS FOR REAL"

November 22, 1978

Mr. Alfred M. Worden
Alfred M. Worden, Inc.
Palm Beach, Florida

Dear Al:

In Houston we talked of the propose golf tournament and I promised I would let you know if I would be able to get down for the event.

Due to a hand injury, my hand is in a cast, and golf is out for a while. I hope to be back in operation by ski season.

It was great to be able to get together at LBJSC with Chris and his gang. It looks as though the Shuttle is for real.

I hope our paths cross again soon.

Sincerely,

Neil A. Armstrong
Professor of Aerospace Engineering & Applied Mechanics

Alfred M. "Al" Worden became a NASA astronaut in April 1966, one who served with the astronaut support crew for Apollo 9 and backup command module pilot for Apollo 12. In the summer of 1971 he voyaged to the Moon as the command module pilot for Apollo 15 along with Commander David R. Scott and James B. Irwin, lunar module pilot. In completing his space flight, Worden logged 295 hours and 11 minutes in space. In 1972 Worden became senior aerospace scientist at NASA Ames Research Center, the following year becoming the chief of the Systems Study Division at Ames. Upon retiring from NASA in 1975 he became president of Maris Worden Aerospace, Inc., before moving on to other corporate positions.

Al died in his sleep on March 18, 2020, at the age of eighty-eight. He was a great personal friend of mine. Not only did he contribute the foreword to Dear Neil Armstrong: Letters to the First Man on the Moon from All Mankind *(2018) but he also served as the chief technical consultant for the 2018 Hollywood film* First Man, *based on my authorized biography of the same title. Like Neil, Al was a very special man, but he was a character and much more gregarious and public than Neil. There was not a person Al ever met who did not regard him as a friend and as a joy to the world.*

FELLOW ASTRONAUTS AND THE WORLD OF FLIGHT

LBJSC stands for the Lyndon B. Johnson Space Center, formerly the Manned Spacecraft Center, renamed in February 1973 by an act of the U.S. Senate in honor of the late U.S. president and Texas native, Lyndon Baines Johnson. Chris is Christopher C. Kraft, NASA flight director during the Mercury, Gemini, and Apollo programs, who became the director of what would become Johnson Space Center in 1972, serving until his retirement from NASA in 1982.

"INFORMATION MAY BE SENSITIVE"

April 5, 1982

Mr. Eugene A. Cernan
Executive Vice President—International
Coral Petroleum
908 Town & Country Blvd.
Houston, TX 77024

Dear Gene:

It was an unexpected pleasure seeing you in Tucson, and I'm only sorry we were both so busy that we didn't have more time for conversation.

I expect to be in Houston for the OTC [Offshore Technology Conference] show in early June, and perhaps we can get together at that time.

We are negotiating with Lanzagorta (Mexico) for a joint venture in rig manufacturing. They already have such an arrangement with Dresser Industries. It would be very helpful to me to have some idea of the nature of their agreement. If you know anyone that might know the details of that situation, I would appreciate your ferreting them out. Our negotiations are still confidential, and the Dresser information may be sensitive, so treat the matter accordingly.

Perhaps we can get a golf game going one of these days. I'm still very beatable.

All the best.

Sincerely,

Neil A. Armstrong

A RELUCTANT ICON: LETTERS TO NEIL ARMSTRONG

It should not be surprising that the two Apollo astronauts (and good friends going back to their days together at Purdue University), Armstrong and Cernan, both spent some time during their post-NASA lives as "oil men," given the years they spent in the Houston area and their status as American heroes. Also like Neil, Cernan became a naval aviator. Part of the third class of NASA astronauts selected in 1963 (the class after Neil's), Cernan became the second American to walk in space, on Gemini IX, and also the last man to walk on the Moon, as commander of Apollo 17. In between he also was the lunar module pilot for Apollo 10. Following a short stint with Coral Petroleum in 1980, he founded the Cernan Corporation, an aerospace and energy consulting firm.

At Coral Petroleum, Inc., Cernan worked for company founder and well-known oil trader David Bay Chalmers Sr.; CPI filed for bankruptcy in 1983, but by then Cernan had been gone for two years. Incidentally, Chalmers belonged to a family with a long history in the international oil business. His uncle, Charles Ulrick Bay (1888–1955) founded Bay Petroleum in 1937; his son David Bay Chalmers Jr. (b. 1953) came to own Bayoil USA, a company heavily involved in oil trading with the Iraqi government throughout the 1990s and early 2000s and which was subject to a 2004 investigation for its part in the U.N.'s Oil-for-Food Programme, from which Chalmers Jr. was convicted of conspiracy to commit wire fraud.

OTC stands for the Offshore Technology Conference, an annual meeting since 1969 where "energy professionals meet to exchange ideas and opinions to advance scientific and technical knowledge for offshore resources and environmental matters."[26]

Lanzagorta refers to Mexican oil technology entrepreneur Emilio Lanzagorta. Today Lanzagorta serves as a vice-president for SL Global Energy, a company founded in 2004 with a multidisciplinary team of industry experts with global experience, executing services in oil well tools operation, process optimization, and strategy development.

With a history dating back into the late nineteenth century, Dresser Industries was the leading manufacturer of industrial and construction equipment and provider of oilfield drilling and other industrial services. In 1950, the company moved its headquarters from Pennsylvania to Dallas, Texas, to be near the center of the nation's major oil and gas fields, where it continued to purchase or otherwise acquire major companies involved in manufacturing overhead cranes, gasoline-dispensing pumps, and heavy equipment for mining and construction. In 1988 it acquired M. W. Kellogg, which was later merged with Kellogg, Brown and Root, and in 1998 merged into Halliburton (later, in 2001, separating from it again).

FELLOW ASTRONAUTS AND THE WORLD OF FLIGHT

Armstrong's request for some "ferreting" out of information by his friend Gene Cernan from Dresser Industries in 1980 may seem to some a little out of character for Neil—not that he was asking Cernan to do anything illegal, but that he was asking him to do something so "sensitive." One may wonder what Neil would have done if Cernan had asked him to do something similar.

"CIVIL SPACE PROGRAM IS WELL ON ITS WAY TO RECOVERY"

July 8, 1987

Mr. Neil Armstrong
P.O. Box 436
Lebanon, OH 45036

Dear Neil:

As promised in NASA's interim report of July 14, 1986, we have delivered the enclosed status report to President Reagan. The report documents a year-long, agency-wide technical and managerial effort to respond to the recommendations of your Commission and to find solutions to the problems, both obvious and subtle, that have come to light as a result of the Challenger accident.

    I believe that the work we have accomplished to date, the procedural and organizational improvements we have put in place, and the program we have developed to return us to safe space flight all merit your favorable consideration. With the next Space Shuttle flight about one year away, it seems appropriate for all America to share in our growing confidence that the civil space program is well on its way to recovery.

Sincerely,

Jim F.
James C. Fletcher
NASA Administrator

A RELUCTANT ICON: LETTERS TO NEIL ARMSTRONG

"Difficult and challenging effort to return to flight status"

August 27, 1987

Dr. James C. Fletcher
Administrator
National Aeronautics and Space Administration
Washington, D.C. 20546

Dear Jim:

Thank you for your letter and the copy of the status report to the President.

I believe NASA's actions, as described in the report, are comprehensive and responsive to the commission's recommendations. I commend you and all the team for the quality and detail of the difficult and challenging effort to return to flight status.

Today is the day of the full scale firing of the SRB. I know that all of us that endured the trauma of 1986 hope that the test goes well.

All the best.

Sincerely,

Neil A. Armstrong

At President Reagan's request, Armstrong served as the vice-chair of the Presidential Commission on the Space Shuttle Challenger Accident, with former secretary of state William P. Rogers serving as chair. As vice-chair of the commission, Armstrong sat ex officio on all subcommittees, spending most of his time working to pin down the causes of the accident itself rather than looking at broader issues related to NASA management and contractor performance. Neil also did some of his own private investigating, seeking information and insights from some of the private contacts he had developed over thirty years in NASA and the aerospace industry. He accepted the final conclusions of the Rogers Commission, notably that the cause of the Challenger *accident was the failure of the pressure seal in the aft field joint of the shuttle's right solid rocket motor due to a faulty design that was unacceptably sensitive to a number of factors, including cold temperature. Neil played the key role in laying out the basis for the manner of thinking that went into the commission's final report.*

FELLOW ASTRONAUTS AND THE WORLD OF FLIGHT

Dr. James C. Fletcher (1919–1991) served two different times as the NASA Administrator, from April 1971 to May 1977 under President Nixon and again from May 1986 to 1989 under President Reagan. He was therefore responsible for both the early planning of the space shuttle program and its recovery and return to flight after the Challenger *accident.*

"HOPE THAT YOU CAN FEEL COMFORTABLE ABOUT HELPING US WITH THIS PROJECT"

January 22, 1997

Neil A. Armstrong
P.O. Box 436
Lebanon, OH 45036

Dear Neil,

Hope all is well with you and that you survived the holidays intact! We had a great family time this year.

Just a note to Howard Benedict's letter about the possibility of an Alan Bean print as a fund raiser for the Scholarship Fund coincident with the grand opening of the Astronaut Hall of Fame. That event will not be a fund raiser, and as you know, we can always use a little extra here and there. I have been encouraged by the participation of the second and third generations and we always appreciate your help.

I hope that you can feel comfortable about helping us with this project.

All the best,

Alan
Alan B. Shepard
P.O. Box 63
Pebble Beach, CA 93953

A RELUCTANT ICON: LETTERS TO NEIL ARMSTRONG

"Favor the elimination of the 'Hall of Fame' characterization"

February 1, 1997

Dear Alan:

Thanks for your note. I have enclosed a copy of the letter I sent Howard.

You are enshrined in a number of Halls of Fame. My own belief is that many of the Halls that I have learned about in the aerospace field are probably unnecessary. I do support the National Aviation Hall of Fame in Dayton, Ohio, because it is so close and also because it is chartered by Congress.

I certainly believe that the enshrinees in the Astronaut Hall of Fame are worthy but that is not the point. The election should somehow be completely independent. In Dayton, the process involves a lot of people and is somewhat cumbersome, but it does produce a result that is unlikely to be accused of self-interest or bias.

My personal opinion is of little import, but I would favor the elimination of the "Hall of Fame" characterization and replacement with a completely different name. That could over time erase any stigmas that might have been planted in the past. The opening of the new building would provide an excellent opportunity to make such a change.

All the best.

Sincerely,

Neil A. Armstrong

"Hall of Fames have become an interesting phenomenon"

February 1, 1997

Mr. Howard Benedict
Executive Director
Astronaut Scholarship Foundation
6225 Vectorspace Blvd.
Titusville, FL 32780

Dear Howard:

Thank you for your letter of 13 January. Hall of Fames have become an interesting phenomenon. The (original?) which I believe is at New York University (old CCNY) required that individuals not be enshrined until something like 25 years after their death. Their election was by a committee which had careful restrictions on any association between nominators and nominees.

Professional Baseball's Hall of Fame in Cooperstown, as you know, elects their members not by players but by sportswriters. The Ladies Professional Golf Association requires electees to win their way into the LPGA Hall of Fame by winning 30 tournaments. Other Halls have varying, but usually independent, methods of selecting their electees. The Astronaut Hall of Fame would be well advised to study the methods of other Halls and review its policy on selection.

With regard to fund raising techniques, I have found that many artists and publishers use a charitable hook to sell their products and services. Charities, by and large, have few scruples about such matters, and are primarily concerned about their own potential income. Charities often approach me about selling my signature attached to a variety of artifacts, but they are selling signatures. Good art does not need signatures for success. I learned that I must have a reasonable policy and must be consistent in its application. I have found it effective, although some charities have not always kept their word. I will stick with it.

Sincerely,

Neil A. Armstrong

The Astronaut Scholarship Foundation was created in 1984 by the six surviving Mercury Seven astronauts—Scott Carpenter, Gordon Cooper, John Glenn, Walter Schirra, Alan Shepard, and Deke Slayton—as well as by Betty Grissom, widow of the seventh astronaut, Virgil "Gus" Grissom; Dr. William Douglas, the Project Mercury flight surgeon; and Henri Landwirth, an Orlando businessman and friend of the astronauts. Since its creation, the ASF has annually awarded merit-based scholarships to undergraduate juniors and seniors pursuing degrees in science, technology, engineering, and math, the goal being to promote the United States as a global leader in those space-related fields for decades to come.

At the same time, the six then-surviving Mercury astronauts organized the Mercury Seven Foundation, with Howard Benedict, a former space reporter with the Associated Press, as its first executive director. Early on, the foundation conceived of an Astronaut Hall of Fame, which ultimately opened in October

1990 at a facility near the Kennedy Space Center Visitor Center owned by the U.S. Space Camp Foundation. (The original Hall of Fame was located just west of the NASA Causeway, next to the Florida branch of Space Camp.) According to the Astronaut Hall of Fame's own publicity, inductees were selected by a "blue ribbon committee" of former NASA officials and flight controllers, historians, journalists, and other space authorities based on their accomplishments in space or their contributions to the advancement of space exploration. Its inaugural class of inductees comprised the Mercury Seven astronauts themselves. In 1993, 13 astronauts from the Gemini and Apollo programs were inducted, including Neil Armstrong. At the 1997 induction, 24 additional astronauts entered the Hall of Fame, including the remaining Gemini and Apollo astronauts as well as astronauts that were part of the Skylab missions. In his letters to Alan Shepard and Howard Benedict, Neil was indirectly criticizing the Hall of Fame for how it had been inducting some of his fellow astronauts and leaving out others. As of 2017, the Hall of Fame had inducted 388 astronauts plus 97 other space notables, for a total of 485 members.

Alan B. Shepard, the first American in space, died on July 21, 1998, at the age of seventy-four, seventeen months after writing this letter to Armstrong.

Howard Benedict died in April 2005 at age seventy-seven. An award-winning aerospace writer for the Associated Press, he covered more than 2,000 missile and rocket launches, including 65 human flights from Shepard's historic flight in May 1961 to the 34th shuttle mission in 1990. He headed the ASF for more than a decade.

In November 1969 astronaut Alan Bean walked on the Moon as part of Apollo 12, then in 1973 commanded the second Skylab crew. He served as an astronaut for eighteen years before retiring to devote the rest of his life to painting, perhaps becoming more famous for his space art than for his time as an astronaut. He died in May 2018 at age eighty-six after a brief illness.

"I will not participate in the selection process"

October 26, 2004

Email to Howard Benedict from Neil Armstrong

Dear Howard:

Thank you for sending your letter and the bios for the candidates for the Astronaut Hall of Fame.

I will not participate in the selection process. I hope that the number of enshrines in the selection is small compared to the outside committee participants.

Sincerely,

Neil Armstrong

"So far the voting has been smooth"

October 28, 2004

Email to Neil Armstrong from Howard Benedict

Dear Neil:

I appreciate your recent comments on the Astronaut Hall of Fame induction process. I recall that you told Al Shepard several times that you were opposed to astronauts selecting astronauts to an Astronaut Hall of Fame. So I understand your reluctance to participate.

Actually, this is the first time that astronauts have been invited to vote on candidates. If you recall, the first three inductions—in 1990, 1993 and 1997—involved enshrining entire groups of astronauts, those who flew in Mercury, Gemini, and then Apollo. You were in the 1993 Gemini group.

Starting in 2001 we went to an outside committee made up of NASA and industry officials, flight directors, the director of the National Air and Space Museum, historians, journalists and others familiar with the program. This committee of 21 selected four inductees in 2001 and 2003 and five in 2004 (there was a tie for 4th place). These 21 are still on the selection committee, but the Astronaut Scholarship Foundation Board of Directors—made up mostly of astronauts—decided this year to open the voting to all living Astronaut Hall of Fame inductees.

The board also decided that a maximum of three astronauts—the highest vote getters—would be enshrined each year.

So far the voting has been smooth, and many deserving shuttle astronauts are receiving recognition by their peers.

Sorry I rambled on so long, but I wanted to give you background on how this year's voting process came about.

Sincerely,

Howard Benedict

"ALWAYS BE PART OF THE AMERICAN EXPERIENCE"

December 12, 2001

Mr. Neil Armstrong
Post Office Box 436
Lebanon, Ohio 45036

Dear Neil:

Congratulations on receiving the National Aeronautic Association's Wright Brothers Memorial Trophy.
 In 1903, on a cold, windy day at Kitty Hawk, North Carolina, the Wright brothers triumphed over the supposed impossible and achieved powered flight. From those humble beginnings, American aviation has made tremendous advances, and now, less than 100 years after the first flight, man has traveled to the Moon and back, and space shuttles orbit our planet.
 Your achievements in aviation and space exploration are part of the amazing history of flight, and your unforgettable words, "One small step for men; one giant leap for mankind," will always be part of the American experience. Your hard work, determination, and commitment to excellence reflect the spirit of our great Nation.
 Best wishes for future success.

Sincerely,

George W. Bush
The White House
Washington, D.C.

George Walker Bush, the forty-third president of the United States, from 2001 to 2009, was himself a pilot, as was his father, George Herbert Walker Bush, the forty-first U.S. president, from 1989 to 1993. George W. Bush flew Convair F-102s in the Texas Air National Guard and in 1972–1973 drilled with the 187th Fighter Wing of the Alabama Air National Guard, though his time as a pilot in both Texas and Alabama later became controversial during his political career when evidence surfaced that he been favorably treated, despite low pilot aptitude test scores and irregular attendance—this was during the Vietnam War—because of his father's position as a member of the U.S. House of Representatives. The flying career of his father, George H. W. Bush, was far more stellar. Enlisting in the U.S. Navy shortly after the attack on Pearl Harbor, Bush became a naval aviator and one of the youngest ensigns in the U.S. Navy. Assigned to a torpedo squadron, he flew in the Battle of the Philippine Sea, one of the largest air battles of World War II. Promoted to lieutenant (junior grade), he piloted a Grumman TBM Avenger that attacked the Japanese installations on Chichijima in the Bonin Islands in early September 1944. During the attack his aircraft was hit by flak, but Bush managed to release his bombs and scored several hits. His engine ablaze, he flew on several miles before he and two other members of his crew bailed (one, whose parachute failed to open, was killed). Bush spent four hours in his life raft until a U.S. submarine came to his rescue. In November 1944, Bush returned to action, participating in operations in the Philippines until his squadron was replaced and sent home to the United States. By the end of 1944 he had flown fifty-eight combat missions for which he received the Distinguished Flying Cross.

Neil had met both Bushes and was well aware in particular of fellow naval aviator George H. W. Bush's experiences in World War II.

"LIFE LESSONS LEARNED FROM SPACE"

January 8, 2004

Neil Armstrong
Cincinnati, OH 45243

Dear Neil,

My apologies for starting the New Year by asking you for help, but that is just what I am doing. As you know, the Astronaut Scholarship Foundation is committed to strengthening America's position in science and technology by awarding scholarships in these fields. Our scholarship fund is made possible through the outstanding efforts of many members of the Astronaut Hall of Fame as well as many others. Last year, the Foundation awarded 17 scholarships of $8,500 each. In 2004 the value of each scholarship will increase to $10,000, and in 2005, three additional scholarships will be awarded for a total of $200,000 annually.

Walt Cunningham came to us with an idea for a book that will help fund our growing scholarship program. He has an agent and the book "packager" who is both experienced with books of short essays and excited about the prospects for the book. The details are described more fully in Walt's enclosed letter. I am excited about Walt's project because he has agreed to contribute his time for the next nine months and all the profits from the book to the Astronaut Scholarship Fund.

I am hopeful that many of our Hall of Fame members will agree to participate in order to maximize the advance from a publisher. We cannot estimate how much the Foundation might realize until a publisher is onboard and you will be under no obligation until there is a signed contract with a major publisher.

I wholeheartedly endorse Walt's proposal and am urging you to support his efforts. It will enable you to make a real contribution to the Foundation without writing a check. Your essay may take a little thought but the logistics are being set up to minimize the imposition on your time. Walt's book of short essays will not restrict any plans you may have for a book of your own and some of you may already have an appropriate essay from your earlier writings. Hopefully, you will agree that the small imposition on your time is more than offset by the opportunity to enrich our scholarship fund by $100,000 or more.

With your help, Astronauts and Friends, Life Lessons Learned from space can be in the stores for the next Christmas season. All we need from you by January 23 is your commitment to contribute an essay. The essay itself will not be needed until much later. Be assured, the Foundation will assist you in any way we can. Howard Benedict and I will be available by telephone if you have any questions regarding this venture.

Your support over the years has helped make the Astronaut Scholarship Foundation the success it is today. I hope you will agree to help us with our newest opportunity.

Sincerely,

James A. Lovell
Chairman, Astronaut Scholarship Foundation
6225 Vectorspace Blvd.
Titusville, FL 32780

FELLOW ASTRONAUTS AND THE WORLD OF FLIGHT

"What we do need now is your commitment"

January 22, 2004

Email to Neil Armstrong from Walt Cunningham

Hi Neil:

I hope you have received the letters from Jim Lovell and me about our project to raise money for the Astronaut Scholarship Foundation. I know you are working on a book yourself and do not like to be drug into projects like this, but I am hopeful that you will contribute one of the 60-70 essays on your experience in space or flying that we need for the book. The essays can be a paragraph, a page, or more. It can be humorous or serious, lighthearted or contemplative story from your remembrances or personal values. The essay would not be needed for several months and you could even use something you have already written. What we do need now is your commitment, which will help us get the best advance from a publisher, all of which will go to the Astronaut Scholarship Foundation.

If you have any questions, feel free to call [cell and home phone numbers withheld] or Howard Benedict at the Foundation [office phone number withheld].

Walt

"Hoping the Apollo 11 35th anniversary would pass unnoticed"

January 23, 2004

Email to James Lovell from Neil Armstrong

Dear Jim:

I was hoping the Apollo 11 35th anniversary would pass unnoticed.

I have channeled several speaking fees to the Astronaut Scholarship Foundation and am confident that they are well spent. Nevertheless, I prefer to make my educational gifts through my own conduits rather than through the ASF. I have been fortunate enough to be able to provide about a million dollars to Purdue including a presidential scholarship which is a full ride all the way through the PhD. Jerry Ross's daughter was the recipient of such a scholarship (but not mine).

I am not enthusiastic about organizations formed to collect money for good causes. I prefer the more efficient direct giving from donor to recipient charity. Nor am I enamored with such organizations knocking on corporate doors for support. If the cause is good enough, the major Foundations (Ford, Gunn, etc.) are a better route.

I have maintained a policy of not accepting fund-raiser awards or giving fund-raiser appearances and speeches. It has occasionally prevented me from accepting invitations I would have enjoyed, but, on the whole, the policy has served me very well over the years.

Incidentally, I am the person who wrote John Travolta to invite him to participate in the kickoff of the Centennial of Flight in December of 2002. As a result of his acceptance at that event a year ago, he has been, as you know, a very high profile participant at aviation events over the past year.

The ASF has knowingly put itself in the position of having to raise funds and leaning on its associates and friends to help it with its obligations. You have been more than generous in your participation and support of their efforts and I commend you for it.

I had many talks and much correspondence with Al [Shepard] during the early days of the Mercury 7 Foundation and the Astronaut Hall of Fame expressing my views and reservations. I thought then, as I do now, that there were better approaches available for charity and recognition than either of those organizations. I told Al that I would not publically express those views but would not be an active participant. So there is a long history and we are where we are.

As you know. I am rarely enthusiastic about public ceremonies and try to resist my involvement to those that are really meaningful. And fund raisers have never qualified.

That's a long winded answer, but I hope that it will provide some insight into my thinking,

I send you my very best my friend.

Neil

The Centennial of Flight kickoff refers to the first major event to be celebrated under the sponsorship of the U.S. Centennial of Flight Commission, which was

created in 1999 to serve as the coordinating body in charge of planning the activities in commemoration of the one hundredth anniversary of the Wright brothers' historic first powered flight on December 17, 1903, at Kitty Hawk, North Carolina. Neil was a member of the First Flight Centennial Advisory Board, appointed as such by the speaker of the House of Representatives. It was the task of the board to offer advice and counsel to the Centennial of Flight Commission, formulate recommendations to Congress and the executive branch on how to best promote national awareness of the Wright brothers' achievement and its significance and heighten the visibility of all of the centennial activities. The National Centennial of Flight Kick-Off was held at the Smithsonian's National Air and Space Museum on December 17, 2002—exactly one year before the one hundredth anniversary of the Wright flight. Armstrong attended as did his specially invited guest, actor/aviator John Travolta. In the coming months through the events of December 17, 2003, Neil was heavily involved in most of the major activities of the Wright centennial.

Jim Lovell became an astronaut at the same time as Armstrong, as part of NASA's Group 2, or New Nine. He flew into space before Neil did, on the mission of Gemini VIII in December 1965, when he and Frank Borman made a flight lasting 330 hours and 35 minutes, America's longest space mission until Skylab in 1973. Lovell then in November 1966 commanded Gemini XII; along with Buzz Aldrin, they completed a four-day, fifty-nine-revolution flight in Earth orbit, highlighted by three extravehicular activities (EVAs) by Aldrin and a manual rendezvous with an Agena rocket. Lovell next served as command module pilot on the epic journey of Apollo 8, in December 1968, the first manned mission out of Earth orbit. Along with mission Commander Frank Borman and lunar module pilot William A. Anders, Lovell became one of the first humans to orbit the Moon. Lovell's fourth and final spaceflight was the ill-fated Apollo 13, in April 1970. Fifty-five hours into the flight and nearly 200,000 miles from Earth, the cryogenic oxygen system in the service module failed. Lovell, the commander, and his crewmen, John L. Swigert Jr. and Fred W. Haise Jr., working closely with Houston ground controllers, converted their lunar module into an effective lifeboat, found ways to conserve both electrical power and water in sufficient supply to ensure their safety and survival while in space, and managed to return safely to earth. In all, Lovell logged a total of 715 hours and 5 minutes (almost 30 days) in his four spaceflights. In the 1995 movie Apollo 13, *Jim Lovell was played by actor Tom Hanks.*

A RELUCTANT ICON: LETTERS TO NEIL ARMSTRONG

"I know that you are reluctant to make public appearances"

January 27, 2004

Email to Neil Armstrong from Jim Lovell

Dear Neil:

Thank you for your prompt response. I know that it was a long shot but as the saying goes "nothing ventured, nothing gained."

I know that you are reluctant to make public appearances but I did not know your thoughts on fund-raisers or your discussion with Al about the Astronaut Scholarship Foundation.

I have tried to keep ASF a visible legacy to those astronauts of the past but I have run out of ideas on how to make it grow. We have tried foundations to no avail. We do get $100,000 a year from Delaware North, the concessionaire at KSC for our contribution in running the Hall of Fame.

Perhaps the ASF should stay at its present size. The endowment can support about 17 scholars at our present level of funding and pay the small staff if the economy doesn't turn sour.

Again, thanks for responding. Best wishes to you and Carol.

Jim

Delaware North is a food service and hospitality company based in Buffalo, New York. The company, which employs 55,000 people worldwide, offers services in the lodging, sporting, airport, gaming and entertainment industries, and generates some $3.2 billion. Along with NASA Kennedy Space Center in Florida, Delaware North Companies Parks & Resorts, a division founded in 1992, operates the primary concessions at Yellowstone National Park, Grand Canyon National Park, and Niagara Falls State Park.

"Unless it is in dire need of money"

February 10, 2004

Email to Walt Cunningham from Neil Armstrong

Hi Walt:

Thanks for your note. I have been on the road for the last couple weeks and have gotten pretty far behind in my mail. I don't mind putting together some kind of piece for a proposed book, although it is my impression that this type of book rarely does well. I recognize that advances are not now, as they once were, early payments against later sales. It is, I gather, generally a guaranteed minimum payment against future sales. What is not clear to me is why ASF would care about advances, unless it is in dire need of money to meet existing obligations.

As you well know, the publishing world is a tough business.

All the best,

Neil

"Your participation will add much to the project"

February 10, 2004

Email to Neil Armstrong from Walt Cunningham

Many thanks, Neil. Your participation will add much to the project. You may already have something that is appropriate or even an excerpt from your new book. I will be in touch with details on the need date after we have the publisher.

That's right, publishing is tough. I do enjoy writing and am looking forward to putting this project together. Astronauts and Friends may be the exception that proves the rule. Adler Books is quite enthusiastic about the prospects but I doubt if it will come close to your new book. How is your book coming?

I believe ASF is doing better than ever since Delaware North took over the Hall of Fame. Next year, the ASF is increasing to 20 $10,000 scholarships and can use all the funds they can get. I am interested in maximizing the advance just in case it is all we ever realize.

Best to you and Carol and hope to see you soon,

Walt

A RELUCTANT ICON: LETTERS TO NEIL ARMSTRONG

"I have learnt to say NO"

February 10, 2004

Email to Walt Cunningham from Neil Armstrong

Hi Walt,

Thanks for your perspective. My biography is being written by Jim Hansen, an Auburn University history professor. He sends me a lot of questions by E-mail but I understand he has a dozen or more chapters in draft, I am trying not to control it in any way. Que sera, sera.

I am writing a book of my own (not an autobiography) but it is proceeding at a slow pace. I just don't have any time of my own to work on it. I have learnt to say NO if I'm ever to get it finished.

Best,

Neil

"It is not the type of fund-raising that I favor"

June 1, 2005

Email to Al Worden from Neil Armstrong

Hello Al,

I was delighted to receive your letter and see that you are keeping busy. Higher education is a worthy goal and I know you are receiving substantial satisfaction in the process.

I share your enthusiasm and am also actively working for higher education, specifically Purdue University, where I am actively involved in their current capital campaign. Gene Cernan and I previously co-chaired a campaign that was successful in raising about $350 million. Our current campaign is over 1 billion toward its 1.5 billion goal.

I have been happy to have been a contributor as well and was fortunate to be able to give a full ride scholarship through doctorate as well as a number of other very satisfying contributions.

While I do not object to the kind of fund-raising that the Astronaut Scholarship Foundation conducts as indicated in your letter, it is not the type of fund-raising that I favor. So I would prefer to help higher education by continuing my activities in the manner that I have found to work well.

I have not written my autobiography. There is a biography being written by James Hansen, a history of technology professor at Auburn University. I understand it may be out late this year.

I send my best and hope our paths cross sometime soon.

Neil

"WE RESPECTFULLY ASK AGAIN THAT YOU SIGN THE VERIFICATION CERTIFICATE"

March 1, 2005

Neil A. Armstrong—Apollo 11 Astronaut
777 Columbus Avenue
Suite E
Lebanon, Ohio 45036

Buzz Aldrin—Apollo 11 Astronaut
10380 Wilshire Blvd.
Suite 703
Los Angeles, California 90024

Dear Astronauts Armstrong and Aldrin,

This letter respectfully requests your assistance on a matter of historical importance, your July 20, 1969 lunar landing and safe return to Earth.

We four, Henry Gawrylowicz, Anthony Liccardi, Frederick Zito, Robert Zuckerman, are retired NASA aerospace engineers who were assigned to Project Apollo. Our duty station was the Resident Apollo Spacecraft Program Office (RASPO) at the Grumman Aerospace Corporation, Bethpage, New York. We were there for 8 years, 1963 to 1970, during the design, construction and test of all of the Lunar Modules (LM) including your famous spaceship, Lunar Module 5.

During your flight to the moon and safe return, you graciously included in your personal package a small negative 4-1/2 inches x 3-1/2 inches which had on it about 100 signatures of personnel from NASA, NAVY, DELCO, GRUMMAN, MIT and other contractor organizations who had worked on the LM's, including Lunar Module 5. That negative was subsequently returned to the originator, Dr. Frederick A. Zito by NASA and is still in his possession. A copy of the negative is attached for your recollection. We are also attaching a copy of your August 15, 1969 letter thanking RASPO personnel for work done on LM-5 and some representative letters we received from NASA senior officials indicating our work on Lunar Module 5.

The significance of the negative is of great personal value and pride to all of the signatories. We therefore ask for your assistance and cooperation in helping to make it an important and significant part of our family records by verifying its authenticity.

We therefore respectfully ask you, LM-5 Astronauts Neil Armstrong and Buzz Aldrin, to sign the attached Verification Certificate confirming that you did take the negative to the Moon's surface on July 20, 1969 and returned it to Earth. If you wish to change it in any way, please make your revisions and we will have it changed. We would be grateful to you for closing this page in history for us.

A copy of your signed Certification Statement will be sent to as many of the signatory personnel or their families as we can locate after these 36 years. They already have a copy of the negative which was distributed to them in 1969. Your signed Certificate Statement will make it a complete historical document for family records of great personal value to all of us.

It is hoped all is going well with both of you and we wish you the best in your future endeavors. We are still very proud of you, your skill, your patriotism and your bravery.

Please address all communications regarding this matter to Dr. Frederick A. Zito, the originator of the negative. Thank you very much, from all of us.

Sincerely,

Henry Gawrylowicz, NASA Vehicle Manager (RASPO)
Lunar Modules, 2, 7, (Apollo 13), 11 (Apollo 17)

Frederick A. Zito, NASA Guidance, Navigation and Control
Systems Engineer
Lunar Module (RASPO)

FELLOW ASTRONAUTS AND THE WORLD OF FLIGHT

Anthony Liccardi, Assistant Manager,
NASA Resident Apollo Spacecraft Program Office (RASPO)

Robert Zuckerman, NASA Structures and Dynamics Systems Engineer Lunar Module (RASPO)

Dear Neil,

Please sign this certificate and return to me in the envelope provided. I will send to Buzz for signing.
  Thanks very much.

Fred Z.

CERTIFICATE OF VERIFICIATION

March 1, 2005

The copy below is of a negative which was taken to the Moon's surface on July 20, 1969 in LUNAR MODULE 5 and returned to Earth on the Apollo 11 flight by NASA Astronauts Neil A. Armstrong and Buzz Aldrin in recognition and appreciation for technical work done on LUNAR MODULE 5 by the signers of the negative from NASA, NAVY, GRUMMAN, DELCO, AND MIT.

NEIL A. ARMSTRONG, NASA ASTRONAUT, APOLLO 11, LUNAR MODULE 5

BUZZ ALDRIN, NASA ASTRONAUT, APOLLO 11, LUNAR MODULE 5

COPY OF NEGATIVE TAKEN TO THE MOON'S SURFACE ON JULY 20, 1969

A RELUCTANT ICON: LETTERS TO NEIL ARMSTRONG

"We have not heard from you"

July 6, 2005

Neil A. Armstrong
Apollo 11 Astronaut
777 Columbus Avenue
Suite E
Lebanon, Ohio 45036

Dear Neil,

In March of this year Buzz Aldrin mailed you a package of information documents. Those documents asked for your assistance in signing a Verification Certificate confirming that the Apollo 11 Lunar Module crew had taken a 3 inch x 4 inch signed negative to the Moon's surface on July 20, 1969 and returned it to Earth. At our request Buzz had signed the Verification Certificate sometime in March 2005 and then sent it to you for your signature. The cover letter request to you and Buzz was signed by four former NASA engineers whose signatures among many others were on the negative and who had worked for 8 years on the Lunar Module at Grumman where they were assigned as part of the resident NASA contingent. They are Henry Gawrylowicz, NASA Vehicle Manager for LM 2, 7, 11, Anthony Liccardi, Assistant Manager, NASA Resident Apollo Office at Grumman, Frederick Zito, Aerospace Engineer for the Lunar Module Guidance, Navigation and Control System, and Robert Zuckerman, NASA Aerospace Engineer for the Lunar Module Structures and Dynamics system.

    We have not heard from you after these several months and do not otherwise know how to contact you. Therefore, in the event the documents may have become misplaced we are sending you a duplicate set. I have taken the liberty of including several other papers and photographs to help establish that this is indeed a legitimate and reasonable request from your NASA colleagues of over 40 years ago.

    We respectfully ask again that you sign the Verification Certificate. It means a lot to us after these many years. It will be cherished as part of our respective family histories and will not be misused or used for commercial purposes. You have our professional and personal word.

Neil, thank you for your consideration in this matter.

Sincerely,

Dr. Frederick A. Zito
Retired NASA Aerospace Engineer
Babylon, NY 11702
Signed on behalf of Messrs. Gawrylowicz, Liccardi, and Zuckerman.

A retired NASA aerospace engineer, Dr. Fred Zito was a technical liaison for the guidance, navigation, and control equipment that was used successfully in the lunar module that landed on the Moon on July 20, 1969, by Apollo 11 crewmates Armstrong and Aldrin. Zito received five commendations from NASA for his work. Besides his NASA work, Dr. Zito spent part of his career at the Department of Energy at the Brookhaven National Laboratory and the Federal Aviation Administration. Having earned two doctorates—a PhD and EdD—and three master's degrees, Dr. Zito taught courses at five colleges and universities. He served in World War II as U.S. Navy commander, was on active duty in the Korean War, and commanded a school of naval officers in the Ready Reserves during the Vietnam War. Zito died in August 2016.

"I did not recall the signed negative"

July 17, 2005

Email to Dr. Frederick Zito from Neil Armstrong

Dear Dr. Zito,

Thank you for your letter of July 6 and your earlier letter which was forwarded to me by my colleague, Dr. Aldrin.

I did not recall the signed negative. Buzz had signed the certificate and I was pleased, because it suggested that he may have carried it in his Personal Preference Kit. But when I checked with him, he said that he had not carried it but assumed that I had.

Although I was confident that I had not carried it, I intended to check to be certain.

Unfortunately, that material was in storage.

I have traveled to the storage location (about 30 miles from my home) a couple of times and removed boxes that I thought might include my Personal Preference Kit list, unfortunately to no avail. I have been hesitant in responding in hopes that I would find the list and I could definitely confirm the result to you.

There was on board, as you may know, another possible location for such an article. Each flight carried a package of material assembled by NASA officials. I think it was called the Official Flight Kit or OFK. This package was the location of flags and other material that NASA used for their promotional purposes. The crew did not, to my knowledge, have any input into or responsibility for the OFK. It is my belief that if the signed negative was carried on Apollo 11, it was carried in the OFK. But I have no knowledge of who was in charge of that or where information regarding it resides within NASA.

I certainly have no doubt of your turning the negative over and them returning it to you. Both Buzz and I have the highest regard and respect for you and all the other Grumman folks with whom we worked. And I would certainly not hesitate to sign the certificate if I was in a position to so certify. But I am not able to sign something that I cannot personally verify.

I know that is not the answer you had hoped for. I will continue to look for the unfound list. I hope (but do not expect) that it will confirm its inclusion in my PPK.

I include my very best wishes,

Sincerely,

Neil Armstrong

Neil did not sign the Grumman engineers' certificate.

"ASTRONAUTS SLATED TO ATTEND THUS FAR"

January 9, 2009

Email to Neil Armstrong from Jim Lovell

Dear Neil:

I don't know what your plans are for this coming Apollo 11 anniversary but, NASA, Kennedy Space Center and the Astronaut Scholarship

Foundation will be celebrating the anniversary of Apollo 11 and the Apollo program July 24, 2009 at the Kennedy Space Center and would like very much if you would attend. The day will be short consisting of a brief general public and NASA employee stage appearance, dinner hosted by the Foundation and perhaps a short press conference if time permits.

ASF is in the process of inviting the Apollo group and so far they have received an overwhelming response. Astronauts slated to attend thus far include me, Buzz, Duke, Haise, Mitchell, Schweickart, Stafford, Worden and Young. Other astronauts who have also responded include Charlie Bolden, John Blaha, Vance Brand, Dan Brandenstein, Bob Crippen, Jon McBride and Brewster Shaw—and I am putting the bite on my Apollo 8 buddies, Bill and Frank. As you can see, it's already turning into quite the gathering; however, this celebration would not be complete without the entire Apollo 11 crew present.

I appreciate you considering this request. Please let me know if you need additional information and if you are able to attend.

Sincerely,

Jim Lovell

"Expect the Apollo 11 crew to jump at the opportunity"

January 9, 2009

Email to Jim Lovell from Neil Armstrong

Hi Jim,

Thanks for your note. I am aware of more than 2 dozen Apollo 11 anniversary events: MIT, Smithsonian, and several other aviation museums, Grumman alums, National Aviation Hall of Fame, Omega watch, 3 US Navy ships, etc., etc., etc. They all expect the Apollo 11 crew to jump at the opportunity to face more crowds, photographers, etc., etc. You know about that stuff as well as anybody.

As you know, Apollo 11 has had 5th, 10th, 15th, 20th, 25th, 30th, and 35th celebrations. They all take time and effort without compensation. In the past, I have limited my participation to the NASA sponsored events or being a principal speaker at fundraisers. Frankly, I don't know what I am going to do this time but I think I just may go to the Gilbert Islands or somewhere more remote.

Because the Apollo Scholarship Foundation is primarily a fundraising organization, I can't see how it would get very high on the list of competitors. I do hope to go to EPNAA and hope I see you there. Give Marilyn a hug for me.

My best,

Neil

EPNAA is the Early and Pioneer Naval Aviators Association, of which both Armstrong and Lovell were members; the association is commonly known as the Golden Eagles. It was formed in 1956 by a group of early naval aviators who were guests of the U.S. Navy on a cruise aboard the USS Forrestal *(CVA-59), an aircraft supercarrier commissioned in 1955 and the first American carrier specifically designed to support jet aircraft. Along with the early aviators who were its members, the Golden Eagles came to include a select number of naval aviators who, during their tour of duty, were pioneers in the development of new concepts, received special citation for unusual, outstanding performance, or who otherwise warranted special consideration. The EPNAA's annual reunion was an event that Armstrong tried to attend every year.*

When Neil refers in the email to the Apollo Scholarship Foundation, he must have meant the Astronaut Scholarship Foundation.

"Inundated with requests"

June 13, 2009

Email to Jim Lovell from Neil Armstrong

Hi Jim,

Apparently you were trapped by your Adler [Planetarium] colleagues to be the person to approach the A-11 crew. Sorry about that.

You are right about being inundated with requests. There have been a lot of them and they are still coming. I had talked to Public Affairs troops in Washington over a year ago suggesting that with all these Apollo anniversaries coming up they should think through how to best orchestrate the whole process. It was really going to be hard on them if they didn't find a way to consolidate some of the stuff. I suggested that an all Apollo crew celebration might be a solution to consider. They agreed to think about it but they were all tied up with the NASA 50th anniversary last fall and didn't get around to think about it until late. I started urging them to get into it early this year, but when your Chicago buddy wouldn't nominate a NASA Administrator, the PR at headquarters couldn't make up their mind on anything.

Jack Dailey asked me last fall if the A-11 crew would do the Glenn Lecture in 09. I told him I had just done one two years ago and it was a lot of work. I wasn't enthusiastic about doing another but I would canvas the crew. Mike said he wasn't eager to do it, but he would go along with whatever the crew wanted. Buzz never answered me. So I told Jack that we did not have a united crew opinion and assumed that would end the matter.

The NASM apparently then went to NASA and arranged for the A-11 crew to do it as a NASA event as part of the 40th. After meeting with NASA we agreed to do it if we could find some reasonable format. They finally proposed that the CDR, CMP, and LMP and Chris Kraft would each do 15 minutes on any subject they wanted. I don't view this as a promising format, but at this point we are stuck with it. So if you go to it (and I won't encourage you), you should not expect too much.

As you would expect, I have not entertained any interviews. I have answered questions by many journalists by e-mail. I have declined many others.

Just so you are aware of how I have been thinking, I established a priority system for ranking the many appearance requests that are coming my way. They are:

1. Official NASA events
2. Events hosted by non NASA government agencies that were involved in Apollo (i.e. Navy recovery ships)
3. Events hosted by companies and organizations that were actively involved in Apollo
4. Events hosted by professional societies whose members were actively involved in Apollo
5. Aerospace Museums holding celebratory events
6. Events hosted by communities that have a substantial Apollo connection
7. Events honoring Apollo that have no close connection.

I didn't mean this to be a rigid rule but as a guideline to help me sort things out. As you can see, the Adler event is in category 5. I have been invited to similar events by Seattle's Museum of Flight, San Diego A&S Museum, Denver Science Museum, Cradle of Aviation Museum on Long Island, the British Science Museum, and the Museum of Transportation in Lucerne. Many of the museums are tying on to an anniversary event with fund raisers. As I have long had a policy of not speaking at fund raisers, I have not accepted any. As you know, I broke my rule for the A-13 event in Seattle and the A-8 event in San Diego, I think, was a fund raiser. In both those cases, I did not let my participation be known before the event so they could not use it as advertising. It's a slippery slope, as you know.

There are a lot of events for mid July. I'm still trying to decide which are absolutely the most important for me to support. I'll put the Adler in the mix and see how it comes out but will not commit for some time.

On another matter, Ron Kaplan of the National Aviation Hall of Fame in Dayton approached me many months ago and asked if I would be the presenter for Ed White's induction in the NAHF. I told him I would think it over. . . . I concluded that . . . I would present it if I could give it to both of Ed's children together. So that's the way it worked out. I am still a bit uncomfortable with it and if you have any thoughts, I would value them.

My best, my friend,

Neil

All three crew members of Apollo 11 and their wives ultimately attended the event planned by NASA in conjunction with the Smithsonian's National Air and Space Museum. It occurred on the evening of July 19, 2009, when Neil and his mates together gave NASM's annual John H. Glenn Lecture in Space History. Mike Collins spoke first, giving brief, casual, witty, and self-disparaging remarks that charmed the audience in the museum's IMAX theatre. Buzz Aldrin spoke second, relying on teleprompters to proclaim his vision of America's future in space through a long series of elaborate PowerPoint slides. Neil closed the event with a lecture he called "Goddard, Governance, and Geophysics," a title that was so academic-sounding that the audience actually chuckled aloud. He said nothing about Apollo 11—and also nothing about himself. What he presented to the surprised but appreciative audience was a scripted lecture on the history of the research into rockets and spaceflight that was the background to NASA's lunar landing program.

The next day (July 20, 2009) Armstrong and his crewmates were guests of President Obama at the White House. Obama hailed the three men as "genuine American heroes" and asserted that "the touchstone for excellence in exploration and discovery is always going to be represented by the men of Apollo 11." Neil always felt it was a great honor to receive a handshake from the president of the United States, as he felt also in this case, but in the coming months, Neil and fellow Apollo legends Jim Lovell would grow increasingly critical of the space policy of the Obama administration and shared their criticisms in front of pertinent congressional committees.

Incidentally, Neil's reference in the above email to Lovell to "your Chicago buddy" was a reference to Barack Obama. Though born in Cleveland, Ohio, Lovell lived most of his post-NASA life in suburban Chicago; and President Obama, after graduating from Columbia University, worked as a community organizer in Chicago, then, after completing law school at Harvard, became a civil rights attorney and an academic who taught constitutional law at the University of Chicago Law School. Obama then represented Chicago's thirteenth district for three terms in the Illinois Senate before his election to the U.S. Senate from Illinois in 2004.

Armstrong knew that Chicago's Adler Planetarium, on the shore of Lake Michigan, was a special place for Lovell. Jim served for many years on Adler's board of advisors and he had his own permanent exhibit at Adler called "Mission Moon." The exhibit tells the story of the Gemini and Apollo missions through Lovell's eyes, featuring artifacts from his personal collection, including one of his NASA flight suits and salvaged items, such as mission manuals, from Apollo 13.

"MIKE SAYS NO TO ALMOST ALL OF THOSE"

January 23, 2009

Email to Neil Armstrong from Pat Collins (Mrs. Michael Collins)

Neil,

Thanks for keeping us up to speed on the various "wannabee" appearances re the 40th. It sometimes seems to me that some organizations are sending requests to each of us individually, if they feel they know one or the other. Mike says NO to almost all of those (supposing that there will be unavoidable "official" things coming up), and so it's good to know where you are coming from. Buzz, as you are aware, keeps accepting things as they suit him or Lois, and I say more power to him, especially as they just run stuff by us, and take no for an answer. It would be fun to get together (6 of us) at some point but meantime, Mike and I and you and Carol seem to keep pretty busy on our own. HAPPY NEW YEAR to you both, and thanks for staying in touch. Mike is showing reluctant signs of being dragged into the 21st century, and has an iMAC, which I will probably spend at least the weekend getting hooked up, and pray that he HIRES a tutor!

Cheers!!

Pat Collins

Boston-born Patricia Finnegan "Pat" Collins met her future husband, astronaut-to-be Michael Collins, in an officers' mess in 1954 while Mike was serving with the twenty-first Fighter-Bomber Wing at Chaumont-Semoutiers Air Base in France and Pat, a trained social worker, was working for the base's air force service club. They married in the summer of 1957, eventually blessed with a daughter, Kate (who became an actress), a second daughter, Ann, and a son, Michael. Pat was extremely intelligent with a sharp wit. She died on April 9, 2014, at age eighty-three.

"Most of these folks act Flabbergasted"

January 23, 2019

Email to Pat Collins from Neil Armstrong

Thanks, Pat.

Most of these folks act Flabbergasted and can't understand why we couldn't covet the opportunity to go to a "gala" in our honor. But then, maybe it is Neil that is all mixed up.

Neil

FELLOW ASTRONAUTS AND THE WORLD OF FLIGHT

"WOULD BE AKIN TO HAVING REMBRANDT PAINT A PORTRAIT"

March 11, 2009

Email to Neil Armstrong from Bert Ulrich, Public Affairs, NASA Headquarters

Hi Neil:

Vanity Fair is very interested in having a photo of your Apollo 11 crew taken by Annie Leibowitz and published for the 40th anniversary of Apollo 11 in its magazine. As you know, Annie Leibowitz is probably the most recognized photographer of the latter 20th century. If we were living in the 16th century this would be akin to having Rembrandt paint a portrait of Henry Hudson (although Hudson was a bit earlier) and having Rembrandt give Henry a copy! It's an honor and of course even more so a great honor for her. At any rate, think about it and let me know soonest whether you are interested. I told her I would get back to them after the weekend. Thanks!

Bert Ulrich

"It's not something I have a particular interest in"

March 11, 2009

Email to Bert Ulrich from Neil Armstrong

Hi Bert:

Thanks for passing it along. It's not something I have a particular interest in, but I wouldn't stand in the way if the rest of the crew wanted to do it. I have had other opportunities to have Ms. L do my portrait but never did one.

Best,

Neil

A RELUCTANT ICON: LETTERS TO NEIL ARMSTRONG

"I am not fond of Vanity Fair"

March 12, 2009

Email to Bert Ulrich from Mike Collins

Bert:

I am not fond of VANITY FAIR. I do see and read most issues. Its fashion ads, which comprise the bulk of the magazine, are frequently grotesque and degenerate and, despite some excellent in-depth articles, I do not like the overall product. I do not think it should be the center piece for a 40th anniversary crew picture. Annie Leibowitz??? God knows I'm no expert, but I think her work is overrated and not in Karsh's league. I hope Neil declines.

Mike Collins

"Just confirm no and I'll turn it off"

March 12, 2009

Email to Neil Armstrong from Bert Ulrich

Hey Neil,

Mike doesn't want to do it. He says it should be moonwalkers only and so I think you are definitely a no on this. Just confirm no and I'll turn it off.

Thanks!

Bert

P.S. Sorry to keep bothering you on this stuff. I have a feeling there will be more things down the road.

"Let's just forget the whole thing"

March 12, 2009

Email to Bert Ulrich from Neil Armstrong

Ok, Bert.

I have never done ANY "moonwalkers only" things, believing that every crewman was mandatory for success. So let's just forget the whole thing.

Neil

"Proust Questionnaire"

April 16, 2009

Email to Neil Armstrong from Bert Ulrich

Hi Neil:

Here is something creative. I know you don't do interviews but I am not sure if this constitutes as one and so I am sending to you anyway. Vanity Fair is interested in having you fill out the "Proust Questionnaire" for the July issue in honor of the 40th anniversary. I don't know if you know about this questionnaire but basically it was found in the papers of the writer Marcel Proust posthumously (probably in the 1920s). Vanity Fair asks writers, artists, historical figures, politicians, etc., to fill out the questionnaire in each issue of the magazine. You only need to pick 20 or 25 of the questions if you are interested. Thanks Neil. Spring is hopefully finally here!

Bert

"I'll pass"

April 16, 2009

Email to Bert Ulrich from Neil Armstrong

Hi Bert,

Vanity Fair is not my thing. I'll pass. Thanks.

Neil

A RELUCTANT ICON: LETTERS TO NEIL ARMSTRONG

"PREFER THAT MY NAME NOT BE INCLUDED"

June 1, 2009

Email to Buzz Aldrin from Neil Armstrong

Hi Buzz,

Thanks for introducing your version and Andy's improvement. I think your proposal is a suggestion which deserves examination. I determined that Purdue's owning my persona would not affect my ability to comment. After thinking about it today, I concluded that I haven't had enough opportunity to think the matter through and, consequently, I would prefer that my name not be included.

I do appreciate your bringing the matter to my attention and send best wishes for the success of your meeting with the Senators.

All the best,

Neil

Andy is Andrew Aldrin, Buzz's son and youngest of three children, who earned a doctorate in political science from UCLA in 1996. Today, Andy Aldrin is director of the Aldrin Space Institute at Florida Institute of Technology, a multidisciplinary institute created to advance commercial space development. Prior to coming to FIT, Andy served as president of a company known as Moon Express, Inc., a commercial spaceflight company that develops scalable single-stage spacecraft capable of reaching the Moon's surface. He also held executive positions in other aerospace industries, at United Launch Alliance and Boeing.

    It is unclear exactly what sort of document Buzz had prepared and sent to Neil for review or what "meeting with Senators" took place in June 2009. But during that month CNN published an op-ed piece by Buzz in which he argued vigorously "Let's Aim for Mars." Buzz writes: "What we truly need is not more Cold War-style competition but a destination in space that offers great rewards for the risks to achieve it. I believe that destination must be homesteading Mars, the first human colony on another world. By refocusing our space program on Mars for America's future, we can restore the sense of wonder and adventure in space exploration that we knew in the summer of 1969. We won the moon race; now it's time for us to live and work on Mars, first on its moons and then on its surface. Exploring and colonizing Mars can bring us new scientific understanding of climate change, of how planet-wide processes can make a warm and wet world into a barren landscape. By exploring and understanding Mars, we may gain key insights into the past and future of our own world. Just as Mars—a desert planet—gives us insights into global climate change on Earth, the promise awaits for bringing back to life portions of the Red Planet through the application of Earth Science to its similar chemistry, possibly reawakening its life-bearing potential."[27]

Given the public comments Neil had made about the present and future of the U.S. space program in favor of a return to the Moon, it is curious that Buzz would be asking Neil to support his views in favor of heading to Mars first.

"YOU CONTINUE TO INSPIRE THE WORLD"

December 9, 2010

Mr. Neil Armstrong
P.O. Box 436
Lebanon, OH 45036

Neil,

I was incredibly honored to receive the Armstrong Award! What a marvelous evening! Thank you for your kind words. You continue to inspire the world.

With my very best wishes,

Sully
C.B. Sullenberger III
Danville, California

On November 12, 2010, Captain Chesley "Sully" Sullenberger, the "Hero of the Hudson," was awarded by Purdue University, his alma mater, with the university's Neil Armstrong Medal of Excellence, making him only the second person to earn the honor. Armstrong, himself a Purdue alumnus, returned to campus to present the medal at a dinner that was part of Purdue's annual President's Council Weekend. The Neil Armstrong Medal of Excellence recognizes individuals who have embodied the same pioneering spirit, determination, and dedication that distinguished Armstrong's own exploration of space and his later roles as a businessman and scholar. Sullenberger piloted the U.S. Airways jetliner that made an emergency landing in the Hudson River on January 15, 2009, shortly after takeoff from New York's LaGuardia Airport, after a collision with a flock of geese took out both engines. Sullenberger stayed on board until all 155 passengers and crew were safely off the plane and into rescue boats. A native of rural Texas who learned to fly as a teenager, Sullenberger completed a master's degree at Purdue following his graduation from the U.S. Air Force Academy in June 1973.

"I AM WORN DOWN BY CEREMONIES"

July 27, 2009

Email to Larry Kuznetz from Neil Armstrong

Dear Mr. Kuznetz,

Thank you for your letter which awaited my return from all the anniversary celebrations. I understand it is your intention to dedicate the quarantine quarters at the LRL. I wasn't certain what that meant. I doubt that it has been selected as a National Historic Landmark or anything of that sort. Perhaps you could help me understand. I wish I had gone through my mail prior to arriving in Houston for the Splashdown Party. We might have been able to complete a ceremony on the 24th. That might have worked out nicely. I must admit, I am worn down by ceremonies.

My best,

Neil

"Would be an inspiration to all who passed through here"

July 27, 2009

Email to Neil Armstrong from Larry Kuznetz

Dear Neil,

First and foremost, thank you for answering my letter in such a timely manner. As you rightly point out, the former Lunar Receiving Laboratory or LRL has not been designated as a historic landmark. However, Building 37 as it is known today, plays a vital role in support of Shuttle, ISS and Constellation operations (it's the home of Space and Life Sciences, Human Countermeasure Programs to micro G; Crew health and training; crew radiation protection, advanced suit and human performance analysis and other key elements). Yet its historical significance to the many NASA and contractor employees involved in these programs, believe it or not, has been lost.

Space and Life Sciences management in concert with John Hirasaki, Bill Carpentier and I (who now work here), therefore thought it would be an inspiration to all who passed through here to be aware of that history and we embarked on a walkthrough to document and record it. Our initial idea was to create special plaques embossed with the Apollo 11 logo for the sleeping quarters, interview rooms and receiving areas of Columbia and the MQF (the draft text of these plaques is enclosed). When we moved from the plaque to the ceremony dedicating stage however, we felt it would be most appropriately done by representative(s) of the prime crew if at all possible. Because this idea came together rather recently, and recognizing that the anniversary celebrations were already overtaxing your schedule, we decided not to intrude upon it with another last minute request. Instead our first thought was to dedicate it sometime during the anniversary period of the quarantine from July 27th to August 9th. If that doesn't work, a delay commensurate with your schedule would be a second welcome choice. As for the ceremony itself, by comparison to what you've been through, we would endeavor to make this a more intimate event, lasting less than an hour and attended by those closest to the crew during LRL operations and a random sampling of current personnel. In closing, if there is any way you can come, we would be most appreciative and grateful of course and adjust the ceremony date accordingly.

Please feel free to enclose a window of acceptable dates if so inclined.

PS. I assume your closing comment at Space Center Houston's Splashdown Party: "that's the way it was, that's the way it is," was a tribute in part to Walter Cronkite. Whether it was or not, its appropriateness and delivery were absolutely priceless.

My very best regards,

Larry Kuznetz
Houston, Texas

A RELUCTANT ICON: LETTERS TO NEIL ARMSTRONG

Dr. Lawrence H. Kuznetz, with advanced degrees from Columbia University and the University of California, Berkeley, was a flight controller for NASA during the Apollo program. Thereafter, Kuznetz worked in the space shuttle program and served as life science experiment manager for the International Space Station (ISS). For the past several years he has taught at Berkeley, with research projects focused on the design of the type of advanced spacesuit that will allow astronauts to work effectively on Mars, a "blue-collar suit" in which astronauts will be able to be "out and about on Mars 7 to 8 hours a day, seven days a week." A private pilot, Kuznetz is also the author of several books, including a book for children about spacesuits. He has made guest appearances on the Tonight Show *and in his spare time plays keyboard in piano bars and on cruise ships.*

Dr. William "Bill" Carpentier was the flight surgeon assigned to the Apollo 11 mission. During the recovery of the astronauts from their command module Columbia *in the Pacific Ocean, Carpentier was in the primary rescue helicopter that ferried the men to the* Hornet. *Both he and NASA engineer John Hirasaki were confined with the astronauts for their twenty-one-day quarantine in the MQF (mobile quarantine facility) and LRL (lunar receiving laboratory).*

"Normally I keep my e-mail in box below 100 messages"

July 29, 2009

Email to Larry Kuznetz from Neil Armstrong

Hello Larry,

Thank you for your e-mail which filled in a lot of gaps in knowledge about the LRL history of more recent years. I certainly think your idea is appropriate in light of the current usage of the building.

I have been immersed in activities related to the 40th anniversary and have fallen further and further behind as a result. Normally, I keep my e-mail in box below 100 messages. I now have about 800 non spam messages awaiting my action or response. Similarly, I have stacks and stacks of unopened mail. And of course, I have all the other normal responsibilities which I have neglected for the past couple of months while I was writing speeches and otherwise preparing for the many commitments that were part of the activities scheduled. There are a number of other 40th anniversary activities to which I am invited over the next several months. I don't know how many I will be able to accept.

FELLOW ASTRONAUTS AND THE WORLD OF FLIGHT

So I would find it very hard to travel to Houston during the next week. After that, I am committed to being in Colorado for about 10 days then back to Ohio for some business commitments. It would help if I had a commitment in Houston in the near future that would allow me to double up. But it's not in the cards right now. If something pops up, I will be certain to notify you.

Sincerely,

Neil

"ONE OF THE MOST SPECIAL DAYS OF MY LIFE"

February 22, 2011

Dear Mr. Armstrong,

This letter is to give you thanks for making February 3rd, 2011 one of the most special days of my life. (The most special day was 69 years ago when my wife said that she would marry me.) I will never forget the kind words you had for me in your address to the people of the U.S. Space and Rocket Center. They will always be with me.

Thank you for coming to Huntsville, and thank you for handing me the award.

Our walks along the rows of students was highly symbolic for me and for the attendees.

I will always remember what you have done for me.

My wife asked me to also convey her thanks and best wishes to you.

Sincerely,

Georg von Tiesenhausen
Huntsville, Alabama

Georg Heinrich Patrick Baron von Tiesenhausen (1914–2018) was a prominent member of the famous rocket team led by Wernher von Braun at NASA Marshall Space Flight Center in Huntsville, Alabama, that designed the Saturn V rocket that took the Apollo missions to the Moon. He was not among the original Germans who came to Huntsville with von Braun in 1950 in the first wave of what was called Operation Paperclip but had rejoined his old team (he worked with von Braun at the Peenemünde facility during World War II) in the late

1950s, in time to be part of the operation that launched the first American satellite and first U.S. astronauts. His chief contributions lay in designing the mobile launch facilities for the Saturn V and the original concept for the lunar rover. Armstrong respected him so much that he agreed to be in Huntsville on February 3, 2011, to present von Tiesenhausen the NASA Lifetime Achievement Award. He lived to be 104 years old, dying in June 2018. A sympathy letter von Tiesenhausen wrote to the Armstrong family following Neil's death in August 2012 is included in this book's final chapter.

"KNOWING THE RESPECT THAT YOU HAD FOR MY FATHER"

[Undated, circa late 2011]

Email Forwarded to Neil Armstrong by the Society of Experimental Test Pilots

"Please forward to Neil Armstrong"

Dear Mr. Armstrong,

Several days ago I received a telephone call from Janice Dunn of the California Space Authority informing me that this group was preparing, or may have already prepared, legislation to submit to Congress to rename the Dryden Flight Research Center after you. Her reasons were: that your name would better reflect the work that is being done at the Center, that Hugh Dryden was involved in aeronautics NOT the space program, and that nobody in Washington knows who Hugh Dryden was.

Knowing the respect that you had for my father, I am certain that you have had no involvement with the California Space Authority's efforts to rename the Center.

However, I am writing to let you know the possible renaming of the Center has greatly upset and hurt the Dryden children, grandchildren and great grandchildren. The Dryden Flight Research Center is the only national recognition given to my father after he had devoted 45 years in service to the nation, many of them with major involvement in the space program.

FELLOW ASTRONAUTS AND THE WORLD OF FLIGHT

Respectfully,

Nancy Dryden Baker

"I have no interest in having my name attached to anything"

[Undated, circa late 2011]

E-mail to Nancy Dryden Baker from Neil Armstrong

Dear Ms. Baker,

Thank you for your letter which has been forwarded to me.

You are correct. I have had, and do have, enormous respect for your father, whom I met a few times, but really never knew personally.

He was an aerodynamics giant in an era when few knew what the word aerodynamics meant. His early work at the Bureau of Standards in the field of high speed and sonic flow was of enormous importance to America's aeronautical progress. His service as the Director of NACA (and my boss) and as Associate Administrator of NASA was one of distinction. I believe you underestimate the regard in which your father is held in Washington. While young politicians who are not knowledgeable about many things abound inside the beltway, there are also many erudite individuals there who hold his memory in great regard.

I fully understand the concern of you and the Dryden family with regard to the effort of some to change the name of the Dryden Center. When I transferred from the Lewis Laboratory (now Glenn Research Center) to NACA at Edwards, it was called the High Speed Flight Station. When I transferred to the space side of NASA in 1962, it was called the Flight Research Center. From time to time suggestions were offered to name the facility after a person. In all candor, I supported the idea of naming it after Howard Lilly who, in 1948, the year after your father joined NACA, became the first NACA pilot to lose his life in the line of duty when the jet engine in the Douglas D-558-1 'Skystreak' in which he was flying, disintegrated shortly after takeoff causing immediate loss of control with no opportunity to escape. That nomination was not approved and, some years later, the very appropriate nomination of your father's name was approved.

A RELUCTANT ICON: LETTERS TO NEIL ARMSTRONG

I have no interest in having my name attached to anything. When my name became well known, requests to use my name for various objectives were ubiquitous. Many used my name without asking. It was clear to me that I should establish a policy which would free me from having to consider, then approve, or reject each request. That policy became one in which I neither encouraged nor prohibited the use of my name on public facilities but did prohibit such usage on any commercial projects. By and large, that policy has served effectively. There exist several hundred schools bearing my name, many streets, and assorted other usages.

For a third of a century, I have not intervened (either in support of or in opposition to) in such matters as a matter of policy. I do not feel that I should break from that non-interference mode. So while I have no desire to have my name on the Edwards facility, I think it is properly the responsibility of others to decide. Additionally, there may exist law, rules, regulations or policies that would affect such decisions and of which I have no knowledge.

I appreciate your letter and send my very best wishes.

Sincerely,

Neil Armstrong

Dr. Hugh L. Dryden was, indeed, one of America's most prominent aeronautical engineers and aerodynamic researchers. Starting in 1920 he headed the aerodynamics section of the National Bureau of Standards, where he studied air pressures on everything from fan and propeller blades to buildings. In 1931 he became a member of the National Advisory Committee for Aeronautics (NACA), which supervised all of the U.S. government's civilian the laboratory work in flight research, and in 1949 he became the NACA's director of research. In that role he helped shape both policies and programs that led to the development of transonic, supersonic, and even hypersonic research, both through advanced wind tunnel investigation and innovative research airplanes, from the record-setting X-1 to the X-15 rocket aircraft. Dryden's leadership was also important to the establishment of vertical and short takeoff and landing aircraft programs and finding solutions to such fundamental problems in space flight as the atmospheric reentry of piloted spacecraft and ballistic missiles. On October 1, 1958, when the NACA became the nucleus of the new National Aeronautics and Space Administration, Dryden became the NASA's deputy administrator, a position he was still holding at the time of his death in 1965. On March 26, 1976, NASA renamed its Flight Research Center at Edwards AFB in California the NASA Hugh L. Dryden Flight Research Center in his honor. This was rescinded by congressional resolution on March 1, 2014, when the center was renamed the Neil A. Armstrong Flight Research Center (H.R. 667, signed into law by President Barack Obama). The new law still paid homage to Dryden by naming the area surrounding the center the Hugh L. Dryden Aeronautical Test Range. Congress had made at least two prior attempts to rename the facility for Armstrong. Each time the effort was led by congressmen from Southern California and from Ohio, Armstrong's home state.

Nancy (Travers) Dryden Baker (b. 1940) died on December 18, 2012, before the Flight Research Center was named after Armstrong. She had two sisters and one brother, all three having died before her own death (one sister died in infancy). Her mother, Mary Libbie (Travers) Dryden, died in 1995.

"EVERYONE IS BACK SAFELY"

August 19, 2011, 7:44:44 AM MDT

E-mail to Carol Armstrong from Neil Armstrong
Subject: Home again

C—

I am in the Delta Sky Club at LaGuardia awaiting my flight home.

Everyone is back safely but I understand the British Consul office in Kabul was hit by insurgents this morning.

My right hearing aid was damaged by my protective helmet so I will stop at the audiology office on my way home to see if they can repair it.

Hope all is well on Palmyra Drive.

Love,

N

"Wait 'till the audiologist asks"

August 19, 2011, 12:28 PM

E-mail to James Hansen from Carol Armstrong
Subject: Neil

Wait 'till the audiologist asks how he damaged his hearing aid.
   Hope your family is well.

Carol
Carol Armstrong

"So Neil was in Kabul . . . ?"

August 19, 2011, 5:40 PM

E-mail to Carol Armstrong from James Hansen
Subject: Neil

So Neil was in Kabul at the time of the attack? Or somewhere else in Afghanistan?

Jim

"Don't know if they knew about the attack"

August 19, 2011, 7:24 PM

E-mail to James Hansen from Carol Armstrong
Subject: Neil

Yes, he was. He returned this afternoon, but I don't know if they knew about the attack until stateside.

Carol

In 2010 and 2011, Armstrong made two trips to visit American troops stationed in the Middle East as part of the Legends of Aerospace tours organized by Morale Entertainment in association with Armed Forces Entertainment. The purpose of the tours was to "lift the spirits of our brave men and women in uniform." Joining Neil were Jim Lovell and Gene Cernan, the commanders of Apollo 13 and Apollo 17. For both tours the itinerary was extensive, with the entourage flying a total of 17,500 air miles, stopping at military bases and hospitals in Germany, Turkey, Kuwait, Saudi Arabia, Qatar, Oman, and Iraq, as well as a couple of U.S. warships positioned in the Persian Gulf. On the trip into the Middle East in August 2011, Armstrong, Lovell, and Cernan were even jetted quietly into Afghanistan for a quick, unscheduled visit to Kabul. While there they met Afghan officers at Camp Eggers, the headquarters of the NATO-led training mission in Afghanistan. The trip had its dangers. On August 6, 2011, an American CH-47D Chinook military helicopter had been shot down by the Taliban, resulting in thirty-eight deaths—thirty Americans and eight Afghans—with no survivors. On August 19, the day the former astronauts were visiting Kabul, militants attacked a British cultural relations agency in a residential neighborhood of Kabul, setting off huge car bomb explosions and killing eight people in a standoff with the police that lasted several hours while agency employees hid inside. The violence was the latest in a series of high-profile attacks in and around Kabul that added a growing sense of unease in the Afghan capital. In the news coverage of the attacks, it was not mentioned that the Apollo astronauts were in the city at the time. It is not clear how close the men were to the attacks, nor is it clear exactly how Armstrong's protective helmet was damaged.

5

THE CORPORATE WORLD

Although Neil's professional life up to 1980 had taken place in the military service, in civilian government employ, and in the halls of academe, over the next three decades he proved himself to be a very adept businessman and corporate director.

One can see signs of his business-mindedness and entrepreneurial spirit when he was a boy in Ohio. Neil's first employment came in 1940 when he was ten—and weighing barely seventy pounds. For ten cents an hour, he cut grass at a cemetery. Later, at Neumeister's Bakery in the town of Upper Sandusky, he stacked loaves of bread and helped make 110 dozen doughnuts a night. He also scraped the giant dough mixer clean: "I probably got the job because of my small size; I could crawl inside the mixing vats at night and clean them out. The greatest fringe benefit for me was getting to eat the ice cream and homemade chocolates."[28] When the family moved to Wapakoneta in 1944, Neil clerked at a grocery store and a hardware store. Later he did chores at a drugstore for forty cents an hour. His parents let him keep all his wages but expected him to save a substantial part of it for college.

An incident from his days in Upper Sandusky demonstrates his superior work ethic and sense of responsibility. Early one morning Neil and two of his fellow Boy Scots, John "Bud" Blackford and Konstantine "Kotcho" Solacoff, set off for Carey, Ohio, ten miles to the north, on their twenty-mile qualifying hike toward a badge required to make Eagle Scout. After eating lunch at windmill-shaped restaurant the boys started back.

According to Kotcho Solacoff, though "fatigue was setting in, Neil kept pushing us to go faster and faster so he could get to work. We told him to go ahead." Neil started "the Boy Scout pace," alternately walking and then running intervals between roadside telephone poles. "By the time we got home," Kotcho recalls, "we were not only exhausted but we had painful cramps in our legs." The next day Kotcho and Bud found that Neil had made it to the bakery on time.[29]

During his high school years after the family had moved to Wapakoneta, Neil hitchhiked or rode his bike (with no fenders) out to the grass airfield outside of town. He helped with what the airplane mechanics called "top cylinder overhauls" (or just "top overhauls," for short). Once he turned sixteen and had his student pilot's license, he could fly an airplane and made a little money from it, by doing what was called "slow time," which coated the valves of the airplane's engine with high-octane gasoline after the top overhauls. As his mother Viola once related, "He was a little grease monkey out there."[30]

In school at Purdue University, Neil also did odd jobs to make extra money, including an early morning route to deliver newspapers. So, he was not unaccustomed to the labor or business sectors of American society.

Throughout the 1970s, when he was teaching aerospace engineering at the University of Cincinnati, Neil had turned down a number of lucrative business offers. He had resisted them all, until January 1979 (nine months before leaving his university post), when he agreed to become a national spokesperson for the Chrysler Corporation. His first TV commercial for the American car manufacturer came during the telecast of that month's Super Bowl XIII, which Neil attended in Miami in the company of Chrysler execs. More TV spots aired the next day as did splashy print ads in fifty U.S. newspaper markets, showing Neil endorsing Chrysler's new five-year, 50,000-mile protection plan. The press asked questions: Why is Armstrong starting to do advertising now, after all this time? And why Chrysler, of all companies? Neil later explained: "In the Chrysler case, they were under severe attack and in financial difficulty, but they had been perhaps the preeminent engineering leader in automotive products in the United States, just very impressive. I was concerned about them and when their head of marketing approached me to take a role that was not just as a public spokesman but also as someone to be involved in their technical decision-making process, I became attracted to that. I visited Detroit, where I talked to Chrysler head Lee Iacocca and other leading company executives. I had a look at the projects they were working on. I got to know some of their people and concluded that it was something I should try. It wasn't an easy decision, because I hadn't done anything like it before. Yet I decided to try, on the basis of a three-year agreement. I loved the engineering aspects of the job, but I didn't think I was very competent in the role as a spokesman. I tried my best, but it wasn't something I was good at. I was always struggling to do it properly."[31]

In the coming months Armstrong forged professional relationships with General Time Corporation and the Bankers Association of America. "The Quartz watch company had built the timer in the lunar module, so that was the connection there—the technology was good. As it turned out, the product quality was not as good as I thought it should be. As for the American Bankers Association, it was not a commercial organization, but rather did an institutional kind of advertising. We made a couple of ads, but it just didn't come together."[32] Armstrong's trial run as a public spokesperson proved to be temporary, but corporate concerns became his primary focus for the rest of his professional life.

As discussed in one of a series of letters in this chapter, simultaneous with leaving the University of Cincinnati, Neil entered into a business partnership with his brother, Dean, and their second cousin Richard Teichgraber, owner of oil industry supplier International Petroleum Services of El Dorado, Kansas. Dean, formerly the head of a General Motors' Delco Remy transmission plant in Anderson, Indiana, became the IPS president; Neil became an IPS partner and the chair of Cardwell International, Ltd., a new subsidiary that made portable drilling rigs, half of them for overseas sale. Neil and his brother stayed involved with IPS/Cardwell for two years, at which time they sold their interests in the company. Dean later bought a Kansas bank.

By 1982, Neil had several different corporate involvements: "I think some people invited me on their boards precisely because I *didn't* have a business background, but I did have a technical background. So I accepted quite a few different board jobs. I turned down a lot more than I accepted."[33]

The very first board on which Armstrong had agreed to serve, in 1972, was Gates Learjet. Neil flew most of the new and experimental developments in the company's line of business jets. In February 1979 he took off in a new Learjet and climbed over the Atlantic Ocean to an altitude of 51,000 feet in a little over twelve minutes, setting new altitude and climb records for business jets.

THE CORPORATE WORLD

In the spring of 1973 Neil joined the board of Cincinnati Gas & Electric. Soon thereafter he became a board member of the Cincinnati-based Taft Broadcasting, with Neil then developing a close friendship with Taft's dynamic CEO and president, Charles S. Mechem Jr. In January 1978 Armstrong joined the board of United Airlines and in 1980 he joined Cleveland's Eaton Corporation, as well as its AIL Systems subsidiary, which made electronic warfare equipment. In 2000 AIL merged with the EDO Corporation, which Neil chaired until his retirement from corporate life in 2002.

In March 1989, three years after he served as vice-chair of the Rogers Commission investigating the fatal demise of space shuttle *Challenger,* Armstrong joined Thiokol, the company that had made the shuttle's solid rocket boosters (SRBs). With Neil's help, Thiokol managed not only to survive but also to grow, in the expanded form of Cordant Technologies, into a manufacturer of solid rocket motors, jet aircraft engine components, and high-performance fastening systems worth some $2.5 billion, with manufacturing facilities throughout the United States, Europe, and Asia. In 2000 Cordant was acquired in a cash deal by Alcoa, Inc., at which time the Thiokol board on which Neil had been serving for eleven years dissolved.

Reluctant to assess the value of any of the corporate contributions he made over the past thirty years, Armstrong only said, "I felt that in most cases I understood the issues and usually then had a view on what was the proper position on that issue. I felt comfortable in the boardroom."[34]

Thanks to his many corporate involvements, Armstrong, for the first time in his life, also made a good deal of money. Besides handsome compensation for his activities as a director, he was also receiving significant stock options and investing his money wisely. By the time he and Janet divorced in 1994, the couple was worth well over $2 million.

Though he never made a show of his philanthropy, Neil was regularly involved in promoting charitable causes, particularly in and around Ohio. In 1973 he headed the state's Easter Seal campaign. From 1978 to 1985 Armstrong was on the board of directors for the Countryside YMCA in Lebanon, Ohio. From 1976 to 1985 he served on the board of the Cincinnati Museum of Natural History, for the last five years as its chair. From 1988 to 1991 he belonged to the President's Executive Council at the University of Cincinnati. Right up to his death in 2012, he actively participated in the Commonwealth Club and Commercial Club of

Cincinnati, having presided over both. In 1992–1993 he sat on the Ohio Commission on Public Service. In 1982 he narrated the "Lincoln Portrait" with the Cincinnati Pops Orchestra. According to Cincinnati Museum of Natural History director Devere Burt: "His name gave us instant credibility. Anywhere you went looking for money, you simply had to present the letterhead, 'Board Chairman Neil A. Armstrong.'"

For his college alma mater, Neil was perhaps the most active. He served on the board of governors for the Purdue University Foundation from 1979 to 1982, on the school's Engineering Visiting Committee from 1990 to 1995, and from 1990 to 1994 he cochaired with Gene Cernan the university's biggest-ever capital fundraiser, Vision 21. Its goal set at a whopping $250 million, the campaign raised $85 million more, setting an American public university fundraising record.

Armstrong was also involved in a few benevolent causes at the national level. From 1975 to 1977 he cochaired, with Jimmy Doolittle, the Charles A. Lindbergh Memorial Fund, which by the fiftieth anniversary of Lindbergh's historic flight, in May 1977, raised over $5 million for an endowment fund supporting young scientists, explorers, and conservationists. In 1977–1978 Neil accepted an appointment to Jimmy Carter's President's Commission on White House Fellowships. And the National Honorary Council's USS *Constitution* Museum Association counted him as a member from 1996 to 2000.

"YOUR NEW COMPANY"

February 5, 1980

Mr. Neil A. Armstrong
Rivendell Farm
1739 No. State Rd. 123
Lebanon, Ohio 45036

Dear Neil:

Surprise, surprise. I never would have guessed that you were about to switch jobs and I want to know a lot more about what you are up to.

Your new company for drilling rigs and assorted accouterments—what is that? Where will you be headquartered—in Dallas I hope.

Please tell me as much about your new company as you can. I am most interested.

We all enjoyed the pleasure of your company at Waddell last year. Let's work to keep our paths crossing.

Best personal regards,

Ken
K. M. Smith
E-Systems
Dallas, Texas 75222

The identity of "Waddell" mentioned in Ken Smith's letter is unknown.

"A challenge to revert to 'honest work'"

March 19, 1980

Mr. K. M. Smith
Executive Vice President
E-Systems, Inc.
Post Office Box 6030
Dallas, Texas 75222

Dear Ken:

My brother (20 years at G.M.) and I had been searching for several years for a proper business to acquire and never could find the right one.

At the beginning of the year, we found a cousin who bought the defunct Cardwell's rights and prints about ten years ago and had been successfully manufacturing the rigs ever since.

We expect to have some fun and hope to make a profit. I know it will be a challenge to revert to "honest work."

Enclosed is a brochure of the product line. It enjoys a good reputation and the business is buzzing.

I plan to keep my office here, but hope to see you again in Dallas or elsewhere.

All the best.

Sincerely,

A RELUCTANT ICON: LETTERS TO NEIL ARMSTRONG

Neil A. Armstrong

The Armstrong cousin who had acquired Cardwell International, Ltd., was Richard Teichgraber. This business partnership is discussed in the introduction to this chapter.

E-Systems of Dallas, Texas, designed, developed, produced, and serviced high technology systems involving surveillance, verification, and aircraft ground-land navigation equipment. The company also developed electronics programs and systems for business, industrial, and nondefense government programs and agencies. In 1995 E-Systems was acquired by Raytheon and renamed Raytheon Intelligence and Information Systems.

"LEAVING THE LAST BOARD MEETING ON AN EJECTION SEAT"

March 14, 1980

Mr. Charles C. Gates
President and Chairman
The Gates Rubber Company
Denver, Colorado 80217

Dear Charlie:

Please accept this belated apology for leaving the last board meeting on an ejection seat. I had no idea it had got so late (time goes fast when you're having fun) and I made it to the airplane (last connection) just as they were closing the door. I am sorry I ran off without notice.

I have been thinking of your kind re-invitation to join the Roundup Ride and have decided to decline. I didn't originally realize that it was more or less a confirming obligation. In reviewing my existing associations, I have concluded that I should not commit to an additional week or so per year. Besides, Janet would divorce me. I do appreciate being asked. I'm convinced it is a great outing.

Janet and I will join Shorty at the Air Force shindig in New York. He seems to feel it is important to the military sales (although I'm not sure what we'll do if we make such a sale). Further I'll be down to the 1000th airplane ceremony in Tucson with Del de Windt. I think EATON will be a fine company to get the number 1000.

THE CORPORATE WORLD

I hope to see you at one or the other, or certainly at the next board meeting.

All the best.

Sincerely,

Neil A. Armstrong

Charles C. (Cassius) Gates (1921–2005) inherited the ownership of the Gates Rubber Company from his father Charles Gates Sr., who had established the company in 1919. By the time Gates Jr. took it over in 1961 the company had become the world's largest non-tire rubber manufacturer. In the following decades, Gates continued to grow, diversify, and acquire other companies, including Learjet (making it Gates Learjet). When the company was sold to London-based Tomkins plc in 1996, it was worth $1.1 billion.

The "Roundup Ride" most likely refers to the Aviation Roundup event held annually at the Minden Tahoe Airport in Carson Valley, Nevada.

Del de Windt was Edward Mandell de Windt (known to friends as Del), for nearly two decades (1969 to 1986) the head of the industrial giant Eaton Corporation based in Cleveland, Ohio. His retirement in April 1986 culminated a forty-five-year association with the company. Under de Windt's direction, Eaton became a highly diversified global power management company with annual sales that grew to more than $3 billion. Key acquisitions during his tenure included the deal in 1970 for Char-Lynn Company, which produced hydraulic motors for agricultural and industrial equipment; the purchase of the Cutler-Hammer electrical business in 1978; and a 1983 joint venture with Sumitomo Heavy Industries in Japan. In addition to his business accomplishments, de Windt was a well-regarded civic leader in Greater Cleveland, his biggest initiative being Cleveland Tomorrow, which became a national model for civic organization with significant engagement from local CEOs working in collaboration with government in addressing the city's economic challenges. Over the course of his career, de Windt was a director of fifteen major corporations and the Federal Reserve Bank of Cleveland. He died at age ninety-one in 2012. Armstrong became a good friend of de Windt, having joined the Eaton board in 1980.

It is not known who Shorty is. It is not John A. Shorty, the public affairs officer for NASA who from 1959 to 1963 worked as the "voice of Mercury control," for he died in December 1979 at the age of fifty-seven.

"WE'RE NOT MAKING ANY MONEY YET"

February 24, 1984

Wing Commander J. M. Henderson, RAF (Ret.)
"Ardmore"
Frensham Vale
Farnham
Surrey GU10 3 HP
England

Dear Jack,

I was delighted to hear from you after so many years, but saddened to know of your "over-stretched elastic." I have no plans to be in the U.K. in October and see little chance of the circumstances changing. Please convey, however, my thanks for the courtesy of the invitation.

After teaching aero engineering for nine years, I am now the chairman of a small computer systems firm specializing in the corporate aviation and charter operator markets. We're not making any money yet, but enjoying the challenge.

I'll keep your kind invitation to visit in mind and hope I get the opportunity to accept some time soon.

All the best.

Sincerely,

Neil A. Armstrong

Born in New Zealand in 1931, Jack Morton Henderson joined the Royal Air Force in 1950, finished pilot's training in 1953, became a staff flight instructor and test pilot, then joined the Royal Aircraft Establishment's Aero Flight unit in 1961. Following extensive training on the Handley Page 115 aircraft, a delta wing research aircraft built to test the low-speed handling characteristics expected for the Anglo-French Concorde SST under development, Henderson played an instrumental role in both its development as well as that of the Hawker-Siddeley P.1127, which evolved into the British Harrier. With RAE Aero Flight, Jack commanded units formulating and flight-testing novel concepts, such as the jet VSTOL, delta wings, slender wings, and electronic flight controls. Besides test-flying a number of advanced British aircraft, Henderson also tested and advised on American, German, Canadian, and French research aircraft, including the Bell

THE CORPORATE WORLD

X-14, Dornier Do.31, and Ryan VZ-3. He served as RAF project manager for the Anglo-French Jaguar aircraft program and planned and controlled the flying displays and all supporting arrangements for the 1968 and 1970 Farnborough International Airshows. He retired as wing commander on medical grounds in the late 1960s and died after prolonged bouts of illness in September 1990. Jack was very much of Neil's generation and, as test pilots, the two men had a great deal in common, staying friends for some thirty-five years.

"AN AVIATION RELATED SETTING WOULD BE ACCEPTABLE"

December 16, 1987

Dear Mr. Armstrong:

In June, American Express started a new print campaign to celebrate the Card's 18th anniversary. I imagine you have seen it, but just in case, I've attached copies. These ads will continue to feature the most accomplished people between then and now in every field of endeavor and we would be most honored if you would like to join this illustrious group sometime in 1988. In addition to the ones you see here we have or are about to photograph:

Beverly Sills
Tom Selleck
I.M. Pei
Billy Kidd
James Earl Jones
Eric Heiden
Hume Cronyn and Jessica Tandy
James Clavell

Most of the people appearing in this campaign would not normally appear in an ad. However, the response has been terrific as they felt it was dignified and not too commercial.

If you are agreeable, the compensation is $50,000 for two years use of the photos. We would like to shoot sometime in 1988 depending on your schedule and that of the photographer, Annie Leibovitz.

So far everyone we have worked with has enjoyed the experience and the press has had a lot of nice things to say about it too.

Needless to say, you must have an American Express Card and have had no competitive conflicts. We are also hoping that you don't have any current endorsements as we are trying to keep this campaign as "non-commercial" as possible.

Please let me know as soon as you can.
Merry Christmas.

Sincerely,

Daisy Sinclair
Vice President
Manager, Casting Department
Ogilvy & Mather Advertising
2 East 40th Street
New York, NY 10017

"Need to be certain we were of the same mind"

February 15, 1988

Dear Ms. Sinclair,

I have taken a good deal of time before responding to your invitation to participate in the American Express print campaign.

I'm persuaded that the artistry of Annie Leibowitz is considerable and it would be a treat to see her in action. Nevertheless, I would need to be certain that we were of the same mind concerning the content and context of the photograph. For example, an aviation related setting would be acceptable but a space related setting would not.

Let me know what you think.

Sincerely,

Neil A. Armstrong

Ultimately Armstrong turned down the offer.

THE CORPORATE WORLD

"GRATEFUL THAT YOU WOULD TAKE TIME TO HELP GE"

February 27, 1990

Dear Neil:

I wanted to thank you for the tremendous assist you gave us with the Soviet Civil Air Ministry and Aeroflot delegation last week.

As it turned out, there was absolutely nothing we could have done that would have impressed our potential customers more than having you present. Your willingness to sign autographs, and your excellent thoughts on American aviation were just right for this group. You even got mention in their news conference.

I am very grateful that you would take time to help GE and its employees in this very significant way. We are all most appreciative.

We'll let you know the final outcome.

Sincerely,

Brian H. Rowe
Senior Vice President
General Electric Company
One Neumann Way
Cincinnati, OH 45215

A world-renowned jet engine pioneer, English-born Brian H. Rowe (1931–2007) led the aviation division at General Electric for three decades, serving from 1979 to 1993 as the division's president and CEO, as well as corporate senior vice-president. During his thirty-eight years with the company, he played a central role in the global leadership position that GE Aviation enjoys today. While running GE Aviation, Rowe launched several new jet engines, including for business and regional jets, for widebody aircraft, and for fighter planes. The CFM56 GE engine designed under his leadership became the most-produced jet engine in commercial aviation.

"THE NUMBERS SHOULD DO THE TALKING"

December 28, 1992

Mr. Russell W. Meyer, Jr.
Chairman and Chief Executive Officer
Cessna Aircraft Company
Wichita, KS 67277

Dear Russ,

Thank you for your gracious invitation following Eaton's aircraft selection decision.

I was delighted to have the opportunity to fly the Citation V. You have a right to be proud of it. It is, indeed, a significant step up from its predecessor. And your crew did a fine job of showing it off.

Given two contenders that could do the job, I, as a director, believed that the numbers should do the talking. I believe the decision was so based. Eaton seems to be very good at financial analysis.
The Citation V does have the superior cabin. Eaton requires a 2 pilot crew and their pilots favor the 31 cockpit for their operation.

All in all, it was an exceptional opportunity for Eaton to upgrade their fleet with either of two fine candidates and favorable economics. Thanks for the invitation to Wichita. I'll certainly look for the opportunity.

I send my best for a great '93.

Sincerely,

Neil A. Armstrong

The Cessna Citation V was a corporate jet (a stretch version of the Citation II) certified for operation in December 1988. According to Neil's letter, Eaton got a great deal in buying the airplane in 1992, at $650,000 (some sold for as much as a million). The maximum speed of the Citation V was 495 mph but likely averaged about 396 mph on its trips considering lower climb, cruise, and descending speeds.

"YOU ARE AN EXCELLENT DIRECTOR"

July 22, 1999

Dear Neil,

Congratulations on the 30th anniversary of your walk on the Moon. I remember where I was and what I was doing when you took "that step" or "that leap." As a nervous flier, I was so proud of your courage.

It has been fun sitting next to you at the USX Board meetings. You are an excellent Director and I am proud to think of you as a friend. I am so happy about the recognition you are receiving this week, an appropriate tribute.

Enclosed is an article in the New York Times of several days ago. I think [William] Safire is brilliant and a great writer. In a way, his article captures a certain aspect of your heroic trip to the moon.

See you soon. *CONGRATULATIONS!*

Sincerely,

Charles R. Lee

P.S. Best to Carol

In the 1990s Charles R. Lee was a director of the USX Corporation, as was Armstrong, with Lee having previously held various financial and management positions in the steel, transportation, and entertainment industries. He served as president and CEO of GTE and later as chair and co-CEO of Verizon Communications, Inc., and director of DIRECTV. Lee serves on a number of other boards, including responsible positions on various boards of his alma mater, Cornell University. From 1986 to 2001, the United States Steel Corporation, more commonly known as U.S. Steel, went by the name USX. Based in Pittsburgh, as of 2016, the company was the world's twenty-fourth-largest steel producer and second-largest domestic producer of steel.

"WE COULD WRITE A BOOK"

September 7, 1999

Mr. Frank A. Olson
Chairman and CEO
The Hertz Corporation
Park Ridge, NJ 07656

Dear Frank,

How nice to hear from you. I realized the years had been passing by when I noted that you used my address from 6 or 7 years ago. But our undaunted postal system challenged rain, snow and dark of night—and delivered.

Enclosed is the picture for your grandson. I can't turn down my grandchildren either!

I love my platinum card, but find it difficult to use because Hertz employees always know I am coming and await me with a variety of requests. United Air Lines is even worse, so I seldom use my United privileges. Would it be possible for you to arrange for me to get a new Hertz card (perhaps just N.A. Armstrong and otherwise unidentified in Hertz files)? If it can't be done, I will understand.

I often remember the chaotic times we shared at United. You were a great boss, but you certainly made the right decision when you declined our invitation to make it permanent. We could write a book . . .

All the best,

Neil A. Armstrong

Armstrong joined the board of directors for United Airlines in January 1978. Olson became the president and CEO of Hertz in 1993, retiring from the position at the end of 1999.

THE CORPORATE WORLD

"SPACE.COM HAS GROWN RAPIDLY"

September 10, 1999

Mr. Neil Armstrong
Chairman
Computing Technology for Aviation
P.O. Box 436
Lebanon, OH 45036

Dear Mr. Armstrong,

I would like to cordially and respectfully invite you to join the Board of Directors of Space.com, Inc. Since we last talked, Space.com has grown rapidly. The response to our website is positive, with traffic running well ahead of projections. We have established bureaus in Houston, Cape Canaveral, and Washington, D.C. We will soon add a bureau in Pasadena to cover JPL. Next, we intend to focus on the Business section and plan to initiate a Kid's section in early November. We've formed a partnership with Analytical Graphics, Inc., which will provide high quality animation, graphics, and software products for our web-based offerings.

Strategic partnership agreements are also being pursued to take advantage of the convergence occurring between media institutions. We have formed strategic relationships with National Geographic Ventures and U.S. Space Camp. Earlier this week, we informed NASA of our intent to propose a cooperative agreement regarding NASA Television. We are also reviewing proposals inviting our participation from space publications, public broadcast projects, and educational undertakings.

Our financial situation is extremely healthy with strong venture capital support from Venrock Associates and Greylock. We are also fortunate to have several unsolicited offers for additional capitalization standing by.

As Chairman and CEO of Space.com, I rely on our Board of Directors to provide guidance, governance, and strategic direction that will enable the company to take advantage of the many opportunities that are before us. We have also established a Board of Advisors, whose members are selected for their expertise in a particular area of space. A complete membership list of our Directors and Advisors is included in the enclosed materials.

We hope you join us as a member of the Board and we are honored to have your consideration. I would be delighted to meet with you in Ohio at your convenience to discuss any questions you might have regarding the plans of Space.com, your responsibilities as a Board member, and compensation. You can reach me at [phone number withheld]. Thank you very much.

Sincerely,

Lou Dobbs
Chairman and CEO
Space.com, Inc.

Space.com launched in New York City on July 20, 1999, the thirtieth anniversary of the Apollo 11 Moon landing, with the goal of covering in an unprecedented fashion the latest discoveries and missions in space. The company was originally founded by news anchor Lou Dobbs and Rich Zahradnick, with Zahradnick, a best-selling novelist with thirty-plus years of journalistic experience working for CNN, Bloomberg News, AOL, and the Fox Business Network, serving as the company's first president, a position that would by early 2000 be filled by astronaut Sally Ride. Along with Armstrong, several other space notables joined the board of Space.com, including Alexei Leonov, Eugene A. Cernan, and Thomas Stafford. In May 2004, the parent company of Space.com changed its name to Imaginova. In 2009 Imaginova sold the Space.com entity (along with some other properties) to Purch, an online publishing company, which in 2018 was sold to Future US, Inc.

"We all have a passion for space"

December 12, 1999

E-mail to Neil Armstrong from Ray A. Rothrock

Dear Neil:

Many thanks for joining Lou, Mitchell, and me for dinner last Thursday in New York. It is a great honor to have met you.

I hope you will seriously consider the opportunity to serve as a member of the board of directors of Space.com. The American space program led by you and others changed America forever. We believe that the internet has too, and that Space.com will further bring the benefits of space through the internet to people everywhere. Not to sound too altruistic, we also believe that there is a substantial financial return for those of us involved in Space.com. This deal, unlike others in which I am involved, has attracted many people, all for different reasons but with a common interest—space. For sure, we all have a passion for space.

If you have any questions or concerns, please do not hesitate to call me. It would be an honor and a pleasure to work with you on this project.

I look forward to hearing from you in the very near future.

Sincerely,

Ray Rothrock

P.S. I would have sent a hard copy, but I don't have your address.

Formerly the chair and CEO of Venrock, an early-stage tech investment partnership originally formed by the Rockefeller family, Ray A. Rothrock in 2014 became the head of RedSeal, Inc., an enterprise cyber security company based in Sunnyvale, California (which had Venrock as one of its founding investors). During his twenty-five-year career at Venrock, Rothrock invested in over four dozen early-stage technology companies. A leader in the venture capital industry, he was elected by his peers as chair of the National Venture Capital Association for 2012–2013. He is an active "angel" investor who serves on the serves on the boards of a number of companies. Rothrock is also the founding director of the Nuclear Innovation Alliance. In 2018, he authored Digital Resilience, *a book that lays out what he believes would be an effective strategy to win today's global cyber war. Rothrock holds a BS degree in nuclear engineering from Texas A&M University, an MS degree in nuclear engineering from MIT, and an MBA from the Harvard Business School.*

It is not clear who Mitchell is.

A RELUCTANT ICON: LETTERS TO NEIL ARMSTRONG

"Wrapping up the year in very good shape"

December 28, 1999

Dear Neil:

Merry Christmas.

Just a quick note to let you know about our progress at Space.com since we talked last. We are now basically over-subscribed on our second round of financing and Space.com will be a $100 million enterprise by the end of January. In addition, we have reached a tentative agreement with NBC as our strategic media partner and this will be announced in 2 to 3 weeks. We are wrapping up the year in very good shape.

I hope your consideration of our invitation is progressing positively, and I wish you much happiness in the New Year.

All the best.

Sincerely,

Lou Dobbs
Chairman/CEO
Space.com, Inc.

"I am delighted to accept"

January 11, 2000

Mr. Lou Dobbs
Chairman and CEO
Space.com, Inc.
1230 Avenue of the Americas
7th Floor
New York, NY 10020

Dear Lou:

Happy New Year! I still can't get used to writing 2-0-0-0.

Thank you for your letter inviting me to join the board. I am delighted to accept. I look forward to working with you in the days ahead. Just let me know if there is anything I need to do.

On another note, when we were together in New York, I mentioned that I was involved in trying to put together a transaction. As you can see from the attached, we got the job done. We are running the traps with the Feds and hope our shareholders will agree that it is a good thing for both companies.

All the best.

Sincerely,

Neil A. Armstrong

"Next round of financing"

January 18, 2000

Mr. Neil A. Armstrong
Cincinnati, Ohio 45423

Dear Neil,

The purpose of this letter is to let you know that Blue Chip Venture Company has proposed to invest in the next round of financing of Space.com.

For your information, attached is a copy of our investment proposal which was sent to Lou Dobbs on December 23, 1999. The proposal is rather unusual—I trust you will enjoy reading the material. Lou and Space.com leadership responded favorably and asked that I join their Board of Directors.

I'm looking forward to assisting and building the Space.com business. I have served on approximately 30 corporate boards over the past decade. However, all of them were on earth—this is my first exposure to the cosmos.

If you have any comments or ideas about Space.com, please feel free to give me a call at [phone number withheld] during the day and [phone number withheld] during evenings and weekends.

Sincerely,

John H. Wyant
Blue Chip Venture Company
1100 Chiquita Center
Cincinnati, OH 45202

A RELUCTANT ICON: LETTERS TO NEIL ARMSTRONG

John H. "Jack" Wyant (b. 1946) founded Blue Chip Venture Company in 1990 and served as its managing director. A graduate of Denison College in Ohio, Wyant had worked in brand management at the Procter & Gamble Company and the Kings Island division of Taft Broadcasting Company. (Armstrong first met him through his own role on the Taft board.) While at Taft, Wyant created Blue Chip Broadcasting, a group that grew to twenty radio stations. During this time he produced live sports and musical television specials, negotiated joint ventures, and in 1980 launched a television network for the Cincinnati Reds, Chicago White Sox, Chicago Blackhawks, Minnesota Twins, and Minnesota North Stars. With Blue Chip's establishment in 1990, Wyant invested $600 million in over 125 companies through five venture funds. He also became a member of Queen City Angels, a Cincinnati investment group with direct investments in approximately 20 private growth companies.

"Thanks for the Option Agreement"

February 15, 2000

Dear Lou,

I bumped into Sally [Ride] at LaGuardia the other day. She was on her way to Houston. She seems excited about all that you are doing and noted that the second round was going well. (I told her I had bumped into Jack Wyant recently, who also seems excited).
    Thanks for the Option Agreement. It is a bit different from most of the options agreements I hold, so I guess I will have to ask you to send me a copy of the "CERTAIN OPTION AGREEMENT" referenced in Section 10(a) of the document so I can understand a few of the specifics like valuation and restrictions on stock sales after exercise, etc.
    I look forward to seeing you next month.
    All the best.

Sincerely,

Neil A. Armstrong

THE CORPORATE WORLD

"Standard form of Indemnification Agreement"

February 18, 2000

Mr. Neil A. Armstrong
Cincinnati, Ohio 45423

Dear Mr. Armstrong:

At Lou's request I am enclosing 2 copies of the standard form of Indemnification Agreement by which Space.com, Inc., indemnifies members of its Board of Directors. Sally Ride has already signed the agreement as President.

I respectfully ask that you countersign both copies of the agreement, keep one for yourself and return the other one to me in the enclosed envelope.

Please call me if you have any questions.

Very truly yours,

Robert Zeller
Vice President, General Counsel
Space.com, Inc.

"Enjoyed my first experience with the Space.com board"

March 11, 2000

Robert Zeller. Esq.
Vice President, General Counsel
Space.com, Inc.
120 W. 45th St., 35th Floor
New York, NY 10036

Dear Robert:

I enjoyed my first experience with the Space.com board and look forward to our next meeting.

Enclosed please find my executed copy of the Non-Qualified Stock Option Agreement. I kept the copy with Lou's original signature.

I would still like to have a copy of the "CERTAIN OPTION AGREEMENT" referenced in Section 10(a) of the document, as time permits.

I have also enclosed an option exercise form for the 2000 shares that vested on February 11 and the 2000 shares that vested today. I have enclosed a personal check although I noted that the document prefers a certified or cashier's check. If you prefer that type of check, I will obtain one when I return to the U.S. in about 10 days.

I have mailed the Indemnification Agreement under separate cover.

All the best.

Sincerely,

Neil A. Armstrong

"MOST EXCITING EVENT OF THE 20TH CENTURY"

February 12, 2001

Mr. Neil A. Armstrong
Cincinnati, Ohio 45423

Dear Neil,

I recently was reading Arthur M. Schlesinger Jr.'s memoir "Life in the 20th Century."

I thought you'd find the following observation to be interesting: "For me, the most exciting event of the 20th Century was the landing of men on the moon, and I surmise that if our historians 500 years from now remember our century for anything, it will be as the century when man first burst terrestrial bonds and began the exploration of space, the ultimate frontier."

Hope all goes well.

Best regards,

John E. Pepper
Chairman of the Board
Procter & Gamble
Cincinnati, Ohio 45302

THE CORPORATE WORLD

"FULL OF HUMOR BUT A DEEP INSIGHT INTO HISTORY"

December 18, 2001

Mr. Neil A. Armstrong
Cincinnati, Ohio 45423

Dear Neil:

Your ability to land on the moon is nothing compared to your ability to write and deliver an outstanding talk. What a great presentation you made Friday evening. Full of humor but a deep insight into history and how it all started.

Like all the other 800 there to honor you I thoroughly enjoyed it.

Best regards,

Dan McKinnon
President
North American Airlines
JFK International Airport
Jamaica, NY 11430

6
CELEBRITIES, STARS, AND NOTABLES

Being the First Man on the Moon made Armstrong himself a celebrity—not that he ever wanted to be one. But there were aspects of his celebrity status that Armstrong did not totally mind.

As a mere astronaut, Armstrong had the chance to meet a lot of celebrities and important people. But the epochal movement of the first Moon landing had launched all three of the Apollo 11 astronauts—but especially Neil—into what society was recognizing more and more as superstar status.

Neil's first major brush with celebrities came shortly after the Apollo mission ended. On August 13, 1969, at the end of a whirlwind day of ticker tape parades and wild celebrations with huge adoring crowds in New York City and Chicago, the Apollo 11 crew and its entourage put on their tuxedos—and their wives their most elegant ball gowns—to attend a state dinner being put on by President Nixon at the posh Century Plaza Hotel in Los Angeles. President Nixon, his wife, Patricia, and their two grown daughters, Julie and Tricia, hosted the astronauts and their wives in their presidential suite prior to joining dinner guests Mamie Eisenhower, widow of the former president; Esther Goddard, widow of rocket pioneer Robert Goddard; chief justice and Mrs. Warren E. Burger; former vice president and Mrs. Hubert H. Humphrey (among the few Democrats invited); Arizona senator and 1964 Republican presidential nominee Barry Goldwater; and current vice president Spiro Agnew and his wife. Government notables filled the high-domed and elegantly chandeliered banquet hall: NASA and other space program officials, more cabinet members than sometimes attended cabinet meetings, governors of forty-four states (including California governor Ronald Reagan), members of the Joint Chiefs of Staff, diplomatic corps members representing eighty-three nations, and a battery of congressional leaders. U.S. and international aviation pioneers were represented by Jimmy Doolittle, the man who had headed the National Advisory Committee for Aeronautics (NACA) when Neil began his government career in 1955, Wernher von Braun, and Willy Messerschmitt. From Hollywood and show business came entertainers Rudy Vallee, Gene Autry, Jimmy Stewart, Bob Hope, Red Skelton, Rosalind Russell, Art Linkletter, and a score of others. Evangelist Reverend Billy Graham was there. Howard Hughes and Charles Lindbergh had been invited, but neither aviator came out of his self-imposed seclusion to attend. Ironically, not a single member of the Kennedy family attended, indebted as was the occasion to the inspiration of former President John Kennedy. (On July 18, the day Apollo 11 approached lunar orbit, Massachusetts senator Edward "Ted" Kennedy, following a party, had plunged off a bridge at Chappaquiddick Island near Martha's Vineyard, an accident that had killed twenty-eight-year-old campaign worker Mary Jo Kopechne.) The tab for the event, with its 1,440 guests, was $43,000-plus. The president himself had approved the menu right down to the dessert, a sphere of dimpled ice cream topped with a tiny American flag.

A few weeks later, on September 9, the astronauts and their wives attended another fancy affair: the Apollo 11 Splashdown Party sponsored by NASA at the Shoreham Hotel in Washington, D.C., preceded by the formal unveiling at the U.S. Post Office Department of the commemorative Moon landing stamp, the ten-cent stamp that Neil and Buzz canceled after they got back into *Columbia* on July 22. The following week they all returned to Washington, where the Apollo 11 crew was to be honored at a midday joint session of Congress. Promptly at noon, the astronauts were led by a bipartisan delegation up to seats on the Speaker's rostrum. Following a long and loud standing ovation, Armstrong stepped first to the microphone and addressed the assembly, saying that the whole venture began in those halls with the Space Act of 1958. Neil then introduced Buzz, followed by Mike, to the great chamber. When their remarks were concluded, Congress stood and gave them a thunderous applause.

On September 29 the astronauts and their wives headed off on their biggest and broadest post–Apollo 11 tour. Called "Giant Step," the goodwill trip would kick off from Houston, then travel to Mexico City, Bogotá, Buenos

Aires, Rio de Janeiro, Grand Canary Island, Madrid, Paris, Amsterdam, Brussels, Oslo, Cologne, Berlin, London, Rome, Belgrade, Ankara, Kinshasa (Congo), Tehran, Bombay, Dacca, Bangkok, Darwin (Australia), Sydney, Guam, Seoul, Tokyo, Honolulu, and back to Houston. On the trip Neil and his crewmates had the chance to meet and speak with various kings and queens, an emperor, a shah, a pope, prime ministers, members of parliament and other high government officials, ambassadors, famous actors and actresses, musicians and singers, star athletes, Nobel laureates in the sciences and the arts, and other notables. With some of these VIPs Neil, himself a VIP, would keep in touch on an occasional friendly basis.

In the following months Armstrong (minus his Apollo 11 crewmates) would go on the road with one of the world's most popular entertainers of the time, comedian and actor Bob Hope. (As a fellow Buckeye, Hope had already gotten to know Neil well by serving as grand marshal for Neil's post–Apollo 11 parade in Wapakoneta on September 6, 1969.) Neil joined Bob Hope's Christmas 1969 USO tour, what had become an annual event to entertain the U.S. and allied troops in Vietnam, with stops along the way in Germany, Italy, Turkey, Taiwan, and Guam. Actresses Teresa Graves, Romy Schneider, and Connie Stevens, Miss World 1969, the Golddiggers showgirls, and Les Brown and His Band of Renown completed the cast. Under Hope's tutelage, Armstrong, decked out in chino pants, a red sport shirt, and a jungle hat, often played the straight man, such as when Hope said to Neil from on stage in front of a throng of excited and appreciative troops: *Neil, your step on the Moon was the second most dangerous of the year.* To which Neil replied: *Who took the most dangerous?* Hope: *The girl who married Tiny Tim.* (Tiny Tim [Herbert Butros Khaury] was a long-haired ukulele player who sang in a high falsetto voice and had become a pop icon by singing "Tiptoe through the Tulips" and having a regular role on the top-rated NBC comedy *Laugh In*. On December 17, 1969, at the age of thirty-seven, Tiny Tim married seventeen-year-old Victoria Budinger—known as Miss Vicki—on NBC's *The Tonight Show Starring Johnny Carson*, with forty million people watching.)

Following his return to the States after the USO tour, Armstrong for the first time experienced the downside of his celebrity status when stories appeared in the gossip columns that he and actress Connie Stevens, one of the stars of the show, had become romantically involved and that after their return, Neil had been spotted in the audience of Stevens's Las Vegas act. In characteristic manner, Neil would never comment on the story.

CELEBRITIES, STARS, AND NOTABLES

As readers will see in this chapter, Neil, though he certainly never sought out any sort of limelight as a celebrity or VIP, not only started receiving letters of congratulation from many notable individuals and VIPs from around the world soon after his successful Moon landing, he was to become acquainted not only with a significant number of people from many different fields of business and industry, but also with a number of well-known stars from the entertainment field. These many friends and contacts were all in addition to the large network of peers and associates he enjoyed from the world of flying.

CHARLES A. LINDBERGH

March 16, 1969

Mr. Neil A. Armstrong
NASA Manned Spacecraft Center
CB/Astronaut Office
Houston, Texas 77058

Dear Mr. Armstrong:

I have been abroad and travelling most of the time since the launching of Apollo 8—Europe, Asia, the Pacific—with very few days at home here in Connecticut between trips, and those days under high pressure I now find that I have not written to thank you for the courtesies you extended during the visit my wife and I made to the NASA base at Cape Kennedy in December.

I do want you to know how much we appreciated your taking the time to show us around the base and to explain some of the details of the launching and the mission. You added greatly to the interest and enjoyment of our visit. Very many thanks.

With best wishes,

Charles A. Lindbergh
Scotts Cove
Darien, Conn.

A RELUCTANT ICON: LETTERS TO NEIL ARMSTRONG

"Invitation to attend the launching of Apollo XI"

June 15, 1969

Mr. Neil A. Armstrong
NASA Manned Spacecraft Center
CB/Astronaut Office
Houston, Texas 77058

Dear Mr. Armstrong:

Weeks of travelling abroad—Asia and Europe—are responsible for this very late reply to your letter. My wife is delighted that you liked her article on the Apollo 8 launching.

   As for your invitation to attend the launching of Apollo XI, I accept with deep appreciation. But please do not think more about this yourself. You have enough on your hands with plans and final training. Personally, I have always tried to avoid all distractions before a difficult flight so I could concentrate on the essential details of the flight itself. I am most anxious to avoid taking your thought in any way from the fantastically interesting mission you are soon to embark on.

   The official invitation to the launching has just arrived, and I will be mailing my acceptance in the morning. I will be watching every portion of your mission with the greatest interest, and look forward to talking to you about it some day after the pressures have dropped down a bit following your return. I am, of course, also deeply grateful to Colonel Aldrin and Lt. Colonel Collins for their part in extending the invitation.

With best wishes,

Charles A. Lindbergh
Scotts Cove
Darien, Conn.

Neil first met Lindbergh along with his wife, Anne, at the launch of Apollo 8 in December 1968. Neil was given the job of helping with touring Lindbergh around and showing him the facilities. The night before the launch, Neil took Lindbergh out to look at the Saturn V, all illuminated with the xenon lights. Neil then invited the Lindberghs to attend the launch of Apollo 11. They quietly attended the event with their son Jon. Lindbergh later called the Moon landing a "fascinating, extraordinary, and beautifully executed mission." However, he refused President Nixon's invitation to accompany him to the USS Hornet *for the astronauts' recovery, later explaining: "My declining was based on the fact that I spent close to a quarter century achieving a position in which I could live, work, and travel under normal conditions." Apollo 11's splashdown would naturally "attract the greatest concentration of publicity in the history of the world."*[35] *After Apollo 11, Neil had the chance to talk with Lindbergh several times. They sat next to each other at the meeting of the Society of Experimental Test Pilots in Los Angeles in late September 1969.*

Charles A. Lindbergh died on August 26, 1974, at age seventy-two from cancer.

THE MONTGOLFIER BROTHERS

[Translated from French at NASA Headquarters]

[Undated, circa August 1, 1969]

To the NASA European Representative

Dear Sir,

We will appreciate it very much if you will transmit to the Astronauts of Apollo 11 our warm congratulations, and convey to them the admiration which we, like the entire world, felt for their so courageous and so extraordinary exploit.

We are particularly overwhelmed and excited by this event because we are descended from the Montgolfier Brothers. By the grace of NASA and its valiant astronauts—to whom we express all our admiration—our family motto, "SIC ITUR AS ASTRA," has now been realized.

In thanking you for forwarding this message, Sir, may we add our sincere regards.

Bernard de Montgolfier
9, rue des Sources
92 Meudon-Bellevue
France

A RELUCTANT ICON: LETTERS TO NEIL ARMSTRONG

"Revere your famous ancestors"

November 21, 1969

Monsieur Bernard de Montgolfier
9, rue des Sources
92 Meudon-Bellevue
France

Dear Sir,

Thank you very much for your letter of best wishes and congratulations on our recent journey to the moon. We are grateful and proud to have participated in the achievement of our national goal of a successful lunar landing—and return. We believe that as the exploration of our universe expands, so will the benefits of all mankind. We hope that the people of earth are now entering a new era of peace and common understanding.

It is certainly a pleasure to hear personally from a member of the famous Montgolfier family. All of us in aviation, of course, revere your famous ancestors and their contributions that started getting us "off the ground." I, particularly, have been a long-time collector of artifacts concerning the history of flight and will value your letter as an important addition to my collection.

Again, many thanks for your thoughtfulness and consideration.

Most sincerely,

Neil A. Armstrong

JOSEPH E. CRONIN

November 21, 1969

Mr. Joseph E. Cronin
President
The American League of Professional Baseball Clubs
520 Boyiston Street
Boston, Massachusetts 02116

Dear Mr. Cronin:

On behalf of the Apollo 11 crew, please accept our very warm thanks and appreciation for sending us each a Lifetime Pass to American League ballparks. We are grateful for your thus saluting our efforts in the exploration of the moon and appreciate your kind thoughtfulness.

Although I am an Astro fan and my 12 year old son is a Dodger fan, I spent some years of my youth in the Cleveland area, and still retain fond memories of those days as an avid Cleveland fan.

I certainly look forward to the opportunity of using the pass.

Best wishes to you on your important contributions to America's great game.

Sincerely,

Neil A. Armstrong

GOVERNOR LESTER MADDOX

November 21, 1969

Honorable Lester Maddox
Governor of the State of Georgia
Atlanta, Georgia

Dear Governor Maddox:

It was a great pleasure to receive your kind letter and the information of my appointment as an honorary Lieutenant Colonel Aide de Camp on your staff. It was certainly most kind of you to thus honor our efforts on the Apollo 11 flight.

I frequently find occasion to visit the great state of Georgia and congratulate you and the state for the outstanding progress that has been evident there in recent years.

I appreciate your invitation for our suggestions that you may find to be of use and certainly will accept that invitation should the occasion arise.

My very best wishes.

Sincerely,

Neil A. Armstrong

A RELUCTANT ICON: LETTERS TO NEIL ARMSTRONG

SID CAESAR

December 17, 1969

Mr. Sid Caesar
Caesar, Balkin, Brown, Inc.
Penthouse
340 North Camden Drive
Beverly Hills, California 90210

Dear Sid,

I was always impressed with a man who pays his debts! I look forward to the chance to swindle you out of another box of cigars.

All the best for a happy holiday season. May 1970 be as good to you as 1969 was for us.

Sincerely,

Neil A. Armstrong

It is not known when or where Armstrong first met Sid Caesar (1922–2014), one of the greatest comedic stars of the 1950s' golden years of television. Caesar became famous for Your Show of Shows, *which first aired in 1950 and lasted four years; it was followed by* Caesar's Hour, *which combined sketches, musical revues, and situation comedy and stayed on the air until 1957. Both shows featured writers who became famous in their own right, including Neil Simon, Carl Reiner, Mel Brooks, and Larry Gelbart. Caesar also appeared in a number of films, including* It's a Mad, Mad, Mad, Mad World *(1963) and* Grease *(1978), the latter as the memorable Coach Vince Calhoun, a tough-talking (while hilarious) high school gym teacher and coach. He died at age 91 in 2014.*

No doubt Armstrong met Caesar in the aftermath of Apollo 11 rather than previous to it. One might think the two men met during the Christmas season 1969 Bob Hope USO tour to entertain the troops in Vietnam, but Sid Caesar was not part of that trip.

CELEBRITIES, STARS, AND NOTABLES

NORMAN MAILER

February 26, 1970

Dear Mr. Armstrong:

I'm writing the identical letter to Buzz Aldrin. I've received word that neither of you wish to be interviewed by me out of a natural suspicion that I would then proceed to analyze you. I'd like to assure you that is not my intent. I have contempt for the idea of a reporter coming to talk to a man for a half-hour or more and then proceeding to analyze his depths. Second of all, it would be difficult for a man who has studied the space program as I have, not to feel admiration for the discipline and heroism of men who proceed to land skillfully upon the moon. That you are now, willy-nilly, historical figures and so will have your characters analyzed for the rest of your lives by good and bad writers is a matter that is entirely outside your power to prevent. One way or another, there is always enough evidence, information, and material about a historical figure so that people who don't even know him, people who don't even meet him, feel free to analyze his nature. That is part of culture and civilization, if you will. And, obviously, I could hardly write a 600 page book about the Apollo 11 journey to the moon without engaging from time to time in analysis and speculation on your characters. But I don't need an interview for that, I never thought I needed an interview for that. In fact, it is usually easier without an interview, because there are no confidences to breach, nor any sense of betraying the good nature of a couple of well shared drinks. No, I wish to interview both of you to talk about the moon and some of the feelings you may have collected now, after the event. It's not crucial to my book—my book will certainly get written without the interview, but it's fitting the book should have it. I've worked as assiduously as any writer I know to portray the space program in its largest not its smallest dimension. Just in the sense that each of you could say that you are among the best pilots in America, so am I one of the best writers in America, and just as no man smaller in ability than yourselves could or should have been entrusted with the landing on the moon, so I present to you, gentlemen, that your exploits shouldn't be written about by any work-a-day journalist nor even by the best of journalists but, in fact, should be approached by writers who have given as much to their professions as you have to yours. I realize your time is still most precious to you and interviews must be almost unendurable in the repetition of their questions. I now say I'm the man who wrote "repetition kills the soul." It would be my pride not to ask you questions which have been asked before. It would also be my pride not to ask you embarrassing, silly, or personal questions. I thought we might just have a couple of drinks and/or dinner and talk. If you wish the entire conversation to be off the record, or if you wish no direct quotes without your permission to see them first, I'm amenable. I'm not interested in trying to trap brave men into indiscretions. Believe me, writing is difficult enough without going in for such indulgences. If this letter gives either of you or both of you cause to reconsider, then I'll be available on most occasions, within a few days' notice, to come down to Houston, or could meet you in New York if you should be passing through here. On this off-chance, let me give you my telephone number: UL-5-8966. I'm not at home much but my secretary, Carolyn Mason, is invariably there on weekdays.

Yours sincerely,

Norman Mailer
Brooklyn, New York

There is no record of Armstrong answering Mailer's letter. Neil was not a fan of Mailer, a liberal political activist, or of his books, which by 1968 included Armies of the Night, *a searing critique of the Vietnam War and a keen firsthand account of the antiwar movement in the United States. On a special assignment from* Life *magazine for the Apollo mission, Mailer (who had majored in engineering sciences at Harvard) spent time at Mission Control in Houston and at the Kennedy Space Center, witnessed the launch of Apollo 11 live, and sat in on the prelaunch interview with Neil and his crewmates on July 5, 1969 (though he asked no questions and had no personal interviews with the astronauts). The book that Mailer mentions in his letter became* Of a Fire on the Moon, *published in the summer of 1970, but long excerpts from the book ("A Fire on the Moon," "The Psychology of Astronauts," and "A Dream of the Future's Face") had already appeared in* Life *starting in August 1969, a few weeks after the Apollo 11 mission ended. In an interview for* First Man, *Neil remembers "browsing" the magazine articles. He did not read Mailer's book.*[36]

CELEBRITIES, STARS, AND NOTABLES

BOB FELLER

October 28, 1970

Mr. Bob Feller
Sheraton-Cleveland Hotel
Cleveland, Ohio

Dear Bob:

It was a pleasure meeting you at the Sheraton during my recent visit to Cleveland. I'll always remember you and the Indians as my favorite player/team combination during my boyhood.

My sons certainly appreciate the two baseballs and look forward to the time when they also will have the opportunity of meeting you.

Thank you again.

Sincerely,

Neil A. Armstrong
Deputy Associate Administrator for Aeronautics
Office of Advanced Research and Technology
NASA
Washington, DC 20546

Bob Feller (1919–2010) was a star pitcher for the American League's Cleveland Indians from 1936 to 1956, with the exception of the 1942–1944 seasons, when Feller served in the U.S. Navy during World War II. Feller pitched a total of three no-hit games and twelve one-hit games and set a modern strikeout record with 18 strikeouts in one game and 348 in one season. He led the American League in victories in four different seasons, finishing with an overall career record of 266 wins and 162 losses. His Cleveland Indians team won two American League pennants during his career (1948, 1954) and one World Series. Neil would have been in college at Purdue both of those years.

THE ANNENBERGS

November 24, 1970

Mr. Neil Armstrong
N.A.S.A.
Washington, D.C. 20546

Dear Neil,

The Ambassador and I were so pleased that you could be with us last Tuesday for our special party for Lord Mountbatten. So many people were thrilled to have an opportunity of meeting you.

As you will recall, during the evening I asked if you would mind autographing a picture for my grandchildren, and as you readily agreed I am taking the liberty of sending you their individual names and addresses. I am sure, as the years pass, your autographed picture will mean more to them than perhaps at this moment while they are so young.

I do hope that when next you are in London we shall have the pleasure of seeing you again.

Warmly,

Lee Annenberg
Mrs. Walter Annenberg
American Embassy
Regent's Park
London N.W.I.
England

Walter Hubert Annenberg (1908–2002) was an American businessman, investor, and owner of The Philadelphia Inquirer, TV Guide, Daily Racing Form, *and* Seventeen *magazine. He was appointed by President Nixon as U.S. Ambassador to the United Kingdom, where he served from 1969 to 1974. His second wife, Leonore "Lee" Cohn Annenberg (1918–2009), was a niece of Harry Cohn, the founder and president of Columbia Pictures. In 1989 the family established the Annenberg Foundation, which provides funding and support to nonprofit organizations in the U.S. and around the world. Notable beneficiaries include the Annenberg/Corporation for Public Broadcasting Project (now Annenberg Learner), which has funded many educational programs on PBS television in the United States; the Annenberg Space for Photography; and the*

CELEBRITIES, STARS, AND NOTABLES

Wallis Annenberg Center for Performing Arts. Walter H. Annenberg headed the Annenberg Foundation until his death in 2002, with Lee, his wife, running it until her death in March 2009.

"*Your personal attention in introducing me to your interesting guests*"

December 9, 1970

Mrs. Walter Annenberg
American Embassy
Regent's Park
London, N.W.I.
England

Dear Lee:

I received your kind note and have forwarded the autographed pictures to your grandchildren as you requested.

I certainly enjoyed your grand party at the residence and very much appreciate your thoughtfulness in inviting me. Your personal attention in introducing me to your interesting guests is certainly very much appreciated.

I expect to return to London shortly after new year and hope that I'll have the opportunity to accept your kind invitation to visit you and the Ambassador again. Please accept my sincere thanks and convey my best wishes to the Ambassador.

Sincerely,

Neil
Neil A. Armstrong
Deputy Associate Administrator for Aeronautics
Office of Advanced Research and Technology
NASA Headquarters
Washington, D.C. 20546

Lord Mountbatten mentioned in Lee Annenberg's letter was Louis Mountbatten (1900–1979), the first Earl Mountbatten (original name Louis Francis Albert Victor Nicholas, prince of Battenberg), British statesman, naval leader, and the last viceroy of India. Entering the Royal Navy in 1913, his career involved major naval commands. During World War II after commanding an aircraft carrier and serving as chief of combined operations, becoming a vice admiral, and de facto joining the chiefs of staff, Mountbatten was appointed supreme allied commander for Southeast Asia, successfully conducting a campaign against Japan that led to the recapture of Burma. As viceroy of India in 1947, he administered the transfer of power from Britain to the newly independent nations of India and Pakistan; as governor-general of India he then helped persuade the Indian princes to merge their states into either India or Pakistan. In 1952 he became commander-in-chief of Britain's Mediterranean fleet and in 1952–1954 and 1959–1965 served as chief of the United Kingdom defense staff and chair of the Chiefs of Staff Committee. In 1965 he became governor of the Isle of Wight. Mountbatten was assassinated in 1979 by members of the Provisional Irish Republican Army, who planted a bomb in his boat.

"Mountbatten Lecture"

July 13, 1971

Honorable Walter Annenberg
American Ambassador
London, England

Dear Mr. Ambassador:

Thank you for your kind letter and endorsement invitations to present the Mountbatten Lecture at the University of Edinburgh. I am pleased to be able to tell you that the National Aeronautics and Space Administration has approved an acceptance of that invitation. In addition, acceptance of the Livingstone Medal of the Royal Scottish Geographical Society and the Freedom of Langholm were also approved. I will be in touch with the various organizations soon advising them of my intentions.

I had the pleasure of visiting England recently to review the collaborative efforts in aeronautical research being conducted by NASA and the Royal Aircraft Establishment under the Ministry of Defence. Although the current Ministry reorganizations create some confusion, you might be interested to know these programs are proceeding most satisfactorily and are of mutual benefit to both the United Kingdom and the United States.

Please accept my sincere thanks for your help and hospitality last November. Your dinner party was both elegant and fun. My wife joins me in sending you and Mrs. Annenberg our sincere best wishes.

Respectfully,

Neil A. Armstrong
Deputy Associate Administrator for Aeronautics
Office of Advanced Research and Technology
NASA Headquarters
Washington, D.C. 20546

On March 10, 1972, Neil presented the Mountbatten Lecture, a prestigious annual event at the University of Edinburgh in Scotland; his lecture was titled "Change and the Space Age." The idea of the Mountbatten lecture series was to bring in each year and an "expert on defence-related matters" to speak to staff, students, and the wider public. After Mountbatten's death in 1979, the title was changed to the "Mountbatten Commemorative Lecture" and as such remained the most honored lecture given at the university each year.[37]

The Livingstone Medal, named after the famed African explorer David Livingstone, has been presented annually since 1901 by the Royal Scottish Geographical Society "for outstanding public service in which geography has played an important part, either by exploration, by administration, or in other directions where its principles have been applied to the benefit of the human race."[38] *Among its recipients prior to Armstrong receiving it in 1971 was polar explorer Ernest Shackleton (1909); Roald Amundsen (1925), the first person to reach the South Pole; pioneering aviator Alan Cobham (1928); Edmund Hillary (1953), the first to reach the summit of Mt. Everest; and the only other American, Commander Robert E. Peary (1903), who on his 1898–1902 expedition set a new "Farthest North" record by reaching Greenland's northernmost point (Cape Morris Jesup) as well the northernmost point of the Western Hemisphere (at the top of Canada's Ellesmere Island).*

The Scottish town of Langholm, on the Scotch-English border (some 18 miles southwest of Glasgow), is the ancestral home of the Armstrong clan. In March 1972, as part of his trip to present the Mountbatten Lecture in Edinburgh, Neil and Janet made a memorable side trip to Langholm, where to the cheers of 8,000 Scots and visiting Englishmen, Neil was named the town's first ever Honorary Freeman. With glorious reception, Neil and his wife rode into town in a horse-drawn carriage, escorted by regimental bagpipers dressed in Armstrong tartan kilts.

During their visit, the Armstrongs also climbed to the top of nearby Gilnockie Tower, built some 500 years ago and for a while the home of Johnnie Armstrong, a notorious border reiver and chieftain who was hanged by a Scottish king, his story romanticized in a novel by Sir Walter Scott. Today, an entire floor of the tower (total of five floors) is devoted to the life of Neil Armstrong. During his visit to Langholm in 1972, Neil told the crowd: "The most difficult place to be recognised is in one's home town. And I consider this now my home town."[39] Langholm has commemorated all of the major Apollo 11 anniversaries and held a fiftieth anniversary event the week of July 20, 2019, at which I was the special guest speaker.

"Armstrong Tartan"

March 8, 1972

Langholm Woolden Crafts Ltd.
Mill Shop
Langholm
Dumfriesshire
Scotland

Dear Professor Armstrong,

May I take this opportunity of extending to you and Mrs. Armstrong a very warm welcome to Langholm on Saturday first.

The reason I write is simply that my Wife and I are going on holiday to Portugal that day, so we will not have the pleasure of sharing with you the Investiture of the Freedom of the Burgh.

The Town Council, to mark the occasion, have asked us supply a back cloth of the Armstrong Tartan, which will be draped from the pulpit during the ceremony.

I should very much like if after the Ceremony you would accept the length of the Tartan, as a symbol of our esteem and a tangible reminder of such an historic occasion, which I feel sure in the years ahead will increase in sentimental value to your family.

The motif on our heading is one which we adopted on that memorable day when your journey to the moon was achieved.

Using the moon as a background, we have super-imposed on it the Monument, which sits on Whita Hill, dominating the town.

I have also left with the Town Clerk, two mounted lengths of Tartan, one of which I should like if you would hand over to your Mother, the other you may wish to retain as a suitable wall decoration and a reminder of your visit to Langholm.

With Best Wishes for Saturday,

Yours sincerely,

A. Stevenson, Chairman

"Pleasant reminder of a most memorable occasion"

March 28, 1972

Mr. A. Stevenson
Chairman
Langholm Woolen Crafts Ltd.
Langholm
Dumfriesshire
Scotland

Dear Mr. Stevenson:

Thank you for your kind letter, which was presented to me upon my arrival in Langholm.

My wife and I enjoyed the ceremony of the presentation of the Freedom of the Burgh immensely. The back cloth of Armstrong Tartan certainly was an attractive addition and I am sorry that your trip away from Scotland prevented your attending the occasion.

We certainly appreciate your thoughtfulness in presenting these lengths of Tartan. I will be most pleased to forward one to my mother, as you requested. Our own length will provide a pleasant reminder of a most memorable occasion.

Best wishes for the continued success of Langholm Woolen Crafts Ltd.

Sincerely,

Neil A. Armstrong
Professor of Aerospace Engineering
University of Cincinnati
Cincinnati, Ohio

A RELUCTANT ICON: LETTERS TO NEIL ARMSTRONG

"Well, the shouting has all died down"

March 30, 1972

Mr. Eddie Armstrong
Town Clerk
Town Hall
Langholm
Dumfriesshire
Scotland

Dear Eddie:

Well, the shouting has all died down and we can all return to our usual responsibilities. The entire event was completed without catastrophe, and with the excellent weather conditions you ordered. I feel it has to be judged as an unqualified success.

The picture I promised will be arriving soon. I'll be mailing it under separate cover. If it arrives damaged, please let me know so that I can order a replacement.

Congratulations are due you on all the aspects of the ceremony. It's clear that all your planning was worthwhile.

My wife's illness was diagnosed as bronchitis, which put her out of commission for a few days, but you'll be pleased to know that she is completely recovered.

Thanks again for all your help. We look forward to seeing you again.

Sincerely,

Neil A. Armstrong
Professor of Aerospace Engineering
University of Cincinnati
Cincinnati, Ohio

"Langholm people"

April 18, 1972

Neil A. Armstrong, Esq.
Professor of Aerospace Engineering
University of Cincinnati

CELEBRITIES, STARS, AND NOTABLES

Cincinnati, Ohio 45221

Dear Neil,

Very many thanks for your letter of 30th March which arrived a few days ago.

I agree the shouting has died down, but, honestly, I do not think Langholm will ever be the same again! You have no idea how much we all enjoyed having you and Janet with us, or how great the impact on our everyday life. Your visit is still a major talking point, and everywhere I go the subject is raised almost as a matter of course. Unquestionably, Langholm's first Freeman seems to have made as a great a mark on 11th March as he did when he landed on the moon—relatively speaking, of course!

On reflection, I think what pleases me most is the fact that you and Janet were instantly and instinctively accepted by our folks, and that the warmth of your welcome was absolutely genuine. As I told you in Edinburgh, Langholm people, although hospitable and friendly, tend to play it cool, but they certainly forgot their reserve immediately you arrived at the old Tollbar, and in the process both thrilled and delighted me. What you sensed and indicated in your address, we, although thousands of miles apart, are of the same kind of people, and I am sure the onlookers recognized this. That you are now very much "one of us" goes without saying, and people are already wondering when you are due back!

From the official point of view, I am glad everything went more or less according to plan, and I only regret the Ducal interference in the afternoon. Had your letter of 22nd February agreeing to spend a little time in Langholm arrived before the Ceremony instead of ten days after, I could have made other arrangements, and, knowing you as I do, I think these would probably have suited both you and Janet better. However, you would at least get some idea of how our nobility operate, and I hope you found your stay at Drumlanrig Castle at least interesting.

I am delighted Janet has now recovered. It looked like flu to me, but bronchitis is perhaps worse, and she certainly has my sympathy as I need to be a martyr to it. In turn, you will be pleased to know Provost Grieve is improving, and, although his sight will be permanently impaired, he should be able to cope fairly easily before long. His nervous system has not quite recovered, but, here again, time is on his side.

I am sorry to say I still not have managed to make suitable arrangements for your "bits and pieces" to be transshipped to you, but I hope to have the problem solved before long. It is not so much the solar value as the intrinsic value which worries me, and I am hoping to make use of diplomatic channels. However, I shall let you know just as soon as I complete my negotiations.

I think that is about all meantime, except to ask yet another favour. Quite apart from many outstanding requests, I am afraid I forgot some of our people on duty, and I should be most grateful if you would kindly autograph the enclosed programmes to help me save face! I shall also be asking you to sign some official photographs, but we can leave that until later.

My warmest regards to you and Janet and to your family.

Yours sincerely,

Eddie

P.S. Our Hon. Treasurer, Miss White, would esteem it a favour if you would kindly also autograph her airmail covers, and the photos.

"Our wonderful visit with you at Drumlanrig"

April 26, 1972

The Duke of Buccleuch
Drumlanrig Castle
Thornhill
Dumfriesshire
Scotland

Dear Walter and Mollie:

Time has passed quickly and Janet and I are once again settled in our home. Janet has fully recovered from her bout with what was diagnosed as bronchitis. Apparently the winds on the ramparts of Edinburgh Castle were too brisk for her Ohio constitution.

We certainly remember warmly our wonderful visit with you at Drumlanrig. Your friendly hospitality made us most comfortable and "at ease" during the weekend. It is certainly a memory we will cherish.

Apparently your forestry made a lasting impression on Janet.

Yesterday she purchased her first hundred trees for planting on our new farm. I think it will be the first group of many.

Please accept our deepest gratitude for your many kindnesses. Should there be anything that we can for you over here, please don't hesitate to let us know.

With all good wishes.

Sincerely,

Neil A. Armstrong
Professor of Aerospace Engineering
University of Cincinnati
Cincinnati, Ohio

The Duke of Buccleuch who hosted the Armstrongs for a weekend visit at Drumlanrig Castle was the ninth Duke of Buccleuch as well as eleventh Duke of Queensberry. During his lifetime (1923–2007) he was Scotland's grandest aristocrat and the largest private landowner in Europe. In 1987 he was ranked as the twentieth richest Briton and Scotland's first or second richest, with his estates valued at upward of 300 million pounds (nearly $400 million). According to his own statements, he came to own 270,000 acres—some 430 square miles—in Scotland and Northamptonshire, with his lands producing each year 127,000 sheep, 13,500 cattle, 18 million liters (4.8 million gallons) of milk, 20,000 tons of cereals, and 50,000 tons of timber. His Christian name was Walter and family name was Scott, clearly indicating that his title traced back to the ducal house of the Scotts of Buccleuch, which had received large grants of lands from King James II of Scotland in the middle of the fifteenth century. Although a political conservative who was regularly the target of Labour Party politicians and other member of the Left, the ninth Duke of Buccleuch was quite broadminded on some topics, including government privatization of the Forestry Commission, which he raged against, and cuts in medical disability benefits, which he also strongly opposed. One of his three principal homes was Drumlanrig Castle, situated on the Queensberry Estate in Dumfries and Galloway, Scotland, sixty-three miles southwest of Edinburgh (and fifty-six miles southeast of Langholm) the castle was constructed between 1679 and 1689 from distinctive pink sandstone and is an example of late seventeenth-century Renaissance architecture. It is a large castle with 120 rooms, 17 turrets, and 4 towers. The Duke of Buccleuch and his wife, Hollie, were avid art collectors, and Drumlanrig Castle featured a magnificent art collection that included Rembrandt's An Old Woman Reading,

Leonardo da Vinci's *Madonna of the Yarnwinder, and several other notable paintings, tapestries, and objects of art. During his visit to the castle Neil surely enjoyed a glass or two of the Duke's special malt whiskey, which he personally supervised in every aspect of its production.*

ERNEST K. GANN

January 18, 1971

Mr. Ernest K. Gann
c/o Simon & Schuster, Inc.
630 Fifth Avenue
New York, New York 10020

Dear Ernie:

Thanks for the copy of your latest. I look forward to spending a quiet evening at home with some antagonists other than my own two sons.

Rumor has it that you and the "Jungmeister" have parted company. Could it be true?

Stop and see us when you get to Washington. Now that I'm a member of the bureaucracy, you should really start worrying about the future of the Country.

All the best in '71.

Sincerely,

Neil
Neil A. Armstrong
Deputy Associate Administrator for Aeronautics
Office of Advanced Research and Technology
NASA Headquarters
Washington, D.C. 20546

Ernest K. Gann (1910–1991) was one of Armstrong's favorite authors, which should be no surprise given that not only was Gann a veteran aviator but a great many of his twenty-one best-selling novels reveal his passion for airplanes and flying. When and where Neil first met Gann is unknown, but it is obvious from this letter that they had met some years earlier, certainly before Apollo 11. In interviews for First Man, *Neil related that he had read several of Gann's*

CELEBRITIES, STARS, AND NOTABLES

books more than once, in particular his novels Island in the Sky *(1944), a fictionalized account of an actual incident in the early years of World War II (in which Gann himself was involved) when a C-47 (DC-3) transport aircraft flying from Greenland to the U.S. is forced to crash-land into the frozen and uncharted regions of North Labrador;* The High and the Mighty *(1953), a dramatic recreation of a real-life trip that Gann flew as a commercial airline pilot for Matson Lines from Honolulu, Hawaii, to Burbank, California; and in particular Gann's autobiographical* Fate is the Hunter *(1961), a classic memoir of early commercial aviation in the U.S. The book referred to by Neil in this letter was* The Antagonists *(1971), a historical novel about the siege of the Masada citadel in Israel by legions of the Roman Empire in AD 73. The novel was adapted as a television miniseries,* Masada, *broadcast in 1981. Gann's other three works mentioned above were all made into major motion pictures.*

The "Jungmeister" mentioned in Neil's letter refers to the Bücker Bü 133 Jungmeister, *an aerobatic plane Gann owned and flew for many years. First built in Germany in 1935, the small single-engine, single-seat biplane served as an advanced trainer for the Luftwaffe into World War II and as airshow aerobatic demonstration aircraft. (At an airshow in Brussels, Belgium, in 1938, a team of three* Jungmeisters *made such an impression on Reichsmarschall Hermann Goering that he ordered a team of nine of the airplanes to be formed.) Armstrong had once flown the plane—the* Jungmeister *remained competitive in international aerobatic competitions into the 1960s—and liked that its controls were light, responsive, and very well balanced, making it an absolute delight to fly. (Incidentally, the original designer of the airplane, Carl Clemens Bücker [1895–1976], was a German aircraft designer who moved to Sweden to manage Svenska Aero AB for a few years after World War I before returning to Germany, bringing back with him brilliant young Swedish aeronautical engineer Anders J. Andersson.)*

HIS ROYAL HIGHNESS, THE PRINCE OF THE NETHERLANDS

January 29, 1971

His Royal Highness
The Prince of the Netherlands
The Hague

A RELUCTANT ICON: LETTERS TO NEIL ARMSTRONG

Your Royal Highness:

I was most honored to receive your kind letter and generous comments concerning my appearance at the World Wildlife Congress in London. I was extremely pleased to be able to contribute to the meeting and felt that I received much more than I presented in information and understanding of the Fund's most important efforts toward wildlife conservation.

You were most kind to ask the Secretary General to place my name on the mailing list for WWF reports and information. I'm very grateful for your thoughtfulness.

Please convey the best wishes of Mrs. Armstrong and myself to Queen Juliana Beatrix. May 1971 be rewarding and enjoyable to each of you.

Respectfully,

Neil A. Armstrong
Deputy Administrator for Aeronautics
Office of Advanced Research and Technology
NASA Headquarters
Washington, D.C. 20546

Neil and Janet Armstrong met Prince Bernhard of the Netherlands (1911–2004) and his wife, Queen Juliana (1909–2004) on October 10, 1969, during their stop in Amsterdam while on the Apollo Giant Step Presidential Goodwill Tour around the world. Queen Juliana authored one of the goodwill messages inscribed on the Apollo 11 silicon chip that Armstrong and Aldrin left on the lunar surface during their EVA (extra vehicular activity). Her message read: "I have great admiration for the skill and perseverance of all those who have contributed to make the first manned flight to the Moon possible. I hope that this achievement will prove of great benefit for the future of mankind."

In his address before the World Wildlife Congress in 1970 Neil told the assembly: "The earth today is an oasis of life in space. It is the only island we know is a suitable home for man. I have a deep sense of the finite significance of our fragility. We are a fragile planet, physically so interdependent. . . We must find ways to protect it. The importance of protecting and saving that home has never been felt more strongly. Protection seems most required, however, not from foreign aggressors or natural calamity, but from its own population." Given these words, and what he wrote to the Prince of the Netherlands, who was at the time the president of the World Wildlife Fund, one might think that Neil became a dedicated member of the WWF and a leading spokesperson for environmental conservation, but, in truth, Neil would have little to do with the organization.

HERMAN WOUK

February 4, 1971

Mr. Neil A. Armstrong
Associate Administrator for Aeronautics
NASA
Washington, D.C. 20546

Dear Neil:

A few of us Washingtonians meet for lunch and conversation once a month. We call it Bill Fay's lunch, after the late Irish ambassador, who was one of the founders of the group. We meet at each other's homes, or at clubs, and invite a guest or two. The food and wine are usually pretty good, but the main thing is the talk. There are ten or eleven of us at the table. We meet at twelve-thirty and break up at two or so.

In March I'm the host, in my home. I've invited my Bohemian Grove camp mate, Glenn Seaborg, and it has occurred to me that you might enjoy the lunch. Two dates are available, March 16th or March 18th. Will you let me know if you can make either one? It's a group of busy men, so we line it up far ahead.

We do most of the talking, unless a guest feels like chiming in. There's nothing to do but come and enjoy yourself, if you're free.

Cordially,

Herman
Herman Wouk
[Street address withheld]
Washington, D.C. 20007

A RELUCTANT ICON: LETTERS TO NEIL ARMSTRONG

"I wish I could accept"

February 18, 1971

Mr. Herman Wouk
[Street address withheld]
Washington, D.C. 20007

Dear Herman:

I was very pleased to receive your kind note and invitation to meet with your conversation circle next month. The opportunity sounds most interesting; I wish I could accept. Unfortunately, my schedule will have me in Paris during that week. I hope, however, that you'll repeat the question at some convenient time in the future.

It was a pleasure to see you again, if only briefly; and I look forward to seeing you again soon—hopefully, under somewhat less hurried circumstances.

Sincerely,

Neil
Neil A. Armstrong
Deputy Associate Administrator for Aeronautics
Office of Advanced Research and Technology
NASA Headquarters
Washington, D.C. 20546

Herman Wouk (1915–2019) was a Pulitzer Prize–winning American novelist best known for his epic war novels—notably The Caine Mutiny *(1951), about a fictional mutiny aboard a U.S. naval vessel during World War II, and his two-volume World War II epic novels,* The Winds of War *(1971) and* War and Remembrance *(1978). Like Neil, Wouk had served in the navy, serving in the Pacific during World War II aboard the destroyer-minesweeper* Zane—*a lot like the USS* Caine *depicted in* The Caine Mutiny—*and Neil very much liked Wouk's books, all of which were meticulously researched. (These two letters amount to the only correspondence between Wouk and Armstrong in Purdue's Neil A. Armstrong papers collection, so it is not known whether Neil ever attended a subsequent Bill Fay lunch.)*

William P. "Bill" Fay was the ambassador (Irish Head of Mission) to the United States from Ireland in the years 1964–1969. He died on September 7, 1969.

CELEBRITIES, STARS, AND NOTABLES

Glenn T. Seaborg (1912–1999) was one of the great chemists of the twentieth century, a Nobel laureate chemist, and discoverer of ten atomic elements including (in 1941) plutonium. For many years he served as associate director-at-large of the Lawrence Berkeley National Laboratory and was University Professor of Chemistry for the University of California. Prior to that, during World War II, he had been a member of the Manhattan Project that designed and built the first atomic bomb. He chaired the Atomic Energy Commission (predecessor to today's U.S. Department of Energy) from 1961 to 1971, under Presidents Kennedy, Johnson, and Nixon. He also presided over both the American Association for the Advancement of Science and the American Chemical Society. A champion for public science education, Seaborg was also a major advocate for nuclear arms control, international cooperation in science, and conservation of natural resources.

LOWELL THOMAS

June 14, 1971

Mr. Lowell Thomas
Honorary President
The Explorers Club
46 East 70th Street
New York, N.Y. 10021

Dear Mr. Thomas:

Thank you for your letter regarding the luncheon honoring the heads of the Reader's Digest Foundation on June 23. Regretfully, I will be out of the country on that day and will be unable to attend.

I certainly enjoyed the Explorer's Club Dinner and was honored to have had the opportunity to meet you and the unbelievable aggregation of distinguished explorers. I look forward to seeing you again in the near future.

Sincere thanks for your thoughtfulness in inviting me to join you at the luncheon.

Sincerely,

Neil A. Armstrong
Deputy Associate Administrator for Aeronautics

Office of Advanced Research and Technology
NASA Headquarters
Washington, D.C. 20546

Lowell J. Thomas (1892–1981) was one of the preeminent American radio commentators of the twentieth century as well as an explorer, a lecturer, an author, and a journalist. He is especially remembered for his association with T. E. Lawrence ("Lawrence of Arabia"), whom he followed into the Arabian Desert in 1917–1918, filing the exclusive story and pictures of the revolt in the desert that helped to make Lawrence (1888–1935) famous. Beginning his radio career at KDKA, a pioneering radio station in Pittsburgh, Thomas in 1930 joined the Columbia Broadcasting System (CBS) as a radio news commentator; during World War II he again worked as a foreign correspondent. He was also a pioneering broadcaster in television, appearing on the first television news broadcast in 1939 and the first daily television program in 1940. For CBS, he covered the U.S. presidential conventions in 1952, 1956, and 1960. His television program High Adventure with Lowell Thomas, *on air in 1957 and 1958, introduced his audience to peoples and customs of remote lands. But his principal medium always remained radio, with his nightly news broadcasts—and signature sign off "So long, until tomorrow"—standing as an American institution for nearly two generations.*

"The Grove is unique"

June 2, 1976

Dear Neil:

I'm sure others have spoken to you about becoming a member of the Bohemian Club in San Francisco, so you will be able to attend the annual fabulous session at the Grove. In fact. Jerry O'Donnell, formerly one of the heads of Pan Am, now lives in the San Francisco area, stopped off to see me over the weekend, and he mentioned this. He said he was sure you had been invited. This came up in connection with some reference to Frank Borman who is about to become a member.

As you may or may not know there are approximately one hundred and thirty different camps in the Grove. I belong to one of the quietest, which includes quite an array of sorta distinguished people. It's usually referred to as the President Hoover camp. Our roster of some thirty men has nearly always included the presidents of Leland-Stanford, the heads of Standard of Indiana, former president of the University of Chicago Larry Kimpton, and some mavericks like myself and a few heads of corporations and so on.

As a non-resident member (as all non-San Franciscans are) I have no "muscle." But, if you haven't already been put up for this, just about the most unusual club in the world, well, I'd like to run interference for you.

Will you be at the Hall of Fame ceremony in mid-July, at Dayton? Jimmy Doolittle, Tom Watson and I will be flying back, leaving the Grove for three days, in order to take part in the Hall of Fame bicentennial event. This time I guess I am scheduled to try and occupy the 7-League Boots you wore at the Dayton ceremony.

The Grove is unique. Let's do something about it.

With a low bow to your bride, and with best wishes.

Yours,

Lowell
Lowell Thomas
Hammersley Hill
Pawling, N.Y. 12564

Hammersley Hill was the name of the handsome redbrick Georgian mansion in which Lowell Thomas and his wife lived on Quaker Hill, New York, a hamlet a mere hour's commute by train from Thomas's office at the RCA Building in Manhattan. Purchased in 1936, the Thomas estate was an absolutely stunning property, much of it majestically situated atop a plateau in fertile Duchess County, with a scenic lake, beautiful rolling farmland, and miles of wooded acreage. When Thomas bought the property, its total value was estimated to be $1.5 million, equivalent today to more than $25 million. And its value only went up. In the late 1930s and early 1940s, the country's most prominent young golf course architect, Robert Trent Jones, built two private courses on the Thomas estate, one of them called the Hammersley Hill course. There is no record of Neil and Janet visiting the Thomases at Hammersley Hill, but it is very possible that they did at some point before Lowell's death in 1981.

The Bohemian Grove is a very private property in northern California belonging to the Bohemian Club, an elite by invitation only, men only social club founded in San Francisco in 1872 by a group of male artists, writers, actors, lawyers, and journalists, all of means and interested in arts and culture. Since its founding, the club has expanded to include politicians, affluent businessmen, and other VIPs. The club is known especially for its annual summer retreat (encampment) at Bohemian Grove in the redwood forest of California's Sonoma County. Notable members over the years have included Henry Kissinger, Walter Cronkite, Richard Nixon, Ronald Reagan, Charles Schwab, Ambrose Bierce, Bret Harte, Mark Twain, and Jack London. Today, the Bohemian Club has grown to a membership of approximately 2,500, including several former U.S. presidents and high-ranking politicians and military officials. The Grove itself covers some 2,700 acres and is virtually inaccessible to the public, lending further mystery to the events and their participants. Activities during the encampment include concerts, theatre, informal lectures (called "Lakeside Talks"), parties, and casual networking and government policy review.

The "7-League Boots" mentioned in Thomas's letter refers to a tradition at the annual banquet of the National Aviation Hall of Fame in Dayton, Ohio, that a special guest is given a pair of boots to wear à la the seven-league boots of European folklore. The story is that the boots allow the person wearing them to take strides of seven leagues per step, resulting in great speed. In such rituals, the boots are supposed to be presented by a magical character to the protagonist to aid in the completion of a significant task.

Neil was enshrined in the National Aviation Hall of Fame in 1979. Lowell Thomas was enshrined posthumously in 1992.

"It's certainly an impressive organization"

June 30, 1976

Mr. Lowell Thomas
Hammersley Hill
Pawling, NY 12564

Dear Lowell:

I was delighted to find your kind letter awaiting one of my infrequent visits to the university during the summer holiday.

I have had the pleasure of visiting the Bohemian Grove during the days when I was a NASA bureaucrat in Washington. It's certainly an impressive organization and the Grove is beautiful and tranquil.

I am most appreciative of your kind offer to help with a nomination to membership. An unsolicited recommendation is a most precious gift. It's not at all clear, however, that I am in a position to attend the meetings.

Perhaps we will have the opportunity to discuss it at the Hall of Fame in Dayton, although I must admit, I have a potential conflict and may not be able to attend.

In any case, I hope to see you soon. Janet joins me in sending all good wishes.

Best,

Neil

Neil never joined the Bohemian Club, but he certainly visited the Bohemian Grove once or twice as a guest during the summer encampment.

"Honorary pallbearer for Lowell"

November 17, 1981

Mrs. Lowell Thomas
Hammersley Hill
Pawling, New York 12564

Dear Mrs. Thomas:

I am genuinely honored to be asked to serve as an honorary pallbearer for Lowell. I was saddened to be prevented from being there, but understand from friends that the service was a fitting tribute to a unique and widely loved and respected man.

Perhaps enough time has passed to allow us to explain my absence. I did get a reservation on the only flight available on the morning of the service. After boarding the airplane in Dayton we were interminably delayed due to a lightning strike on the computer at the Indianapolis Flight Service Station. After two hours on the ground, I had missed not only the service, but the opportunity to pay my respects.

I realize that this is most unimportant to you but I felt that I owed you the explanation. I certainly wanted to be there.

I send my heartfelt hope for your good spirits and look forward to our paths crossing some day.

Sincerely,

Neil A. Armstrong

P.S. No reply needed or expected.

Lowell Thomas died on August 29, 1981.

LEN DAWSON

July 8, 1971

Mr. Len Dawson
[Street address withheld]
North Kansas City, Missouri 64116

Dear Len:

It was a surprise and singular pleasure to receive the autographed copy of your book with the kind note. The book was turned over to my son Rick. In addition to being a promising athlete, he has accumulated a sizeable library of sports biographies. He was, of course, overjoyed to add your volume to his collection.

 I hope I can get back to Lafayette sometime this fall and see a game. Bob DeMoss deserves a better season this year than last and I certainly wish him well.

 May your own career continue to sparkle, as it deserves. I look forward to the opportunity of adding my own cheers of enthusiasm from the sidelines soon.

 Best personal regards.

Sincerely,

Neil
Neil A. Armstrong
Deputy Administrator for Aeronautics
Office of Advanced Research and Technology
NASA Headquarters
Washington, D.C. 20546

Len Dawson (b. 1935) is best known for being the All-Pro quarterback of the NFL's Kansas City Chiefs, leading the Chiefs to a victory in Super Bowl IV in 1970 over the Minnesota Vikings. But Neil knew Dawson much better as the quarterback of his alma mater, Purdue University, from 1954 to 1957 (the first two of those years when Armstrong was still a student at Purdue). During three seasons with the Boilermakers, Dawson threw for over 3,000 yards, leading the Big Ten Conference each time. As a sophomore in 1954 he was the NCAA leader in pass efficiency while also playing defense and serving as the Boilermaker kicker. He was later inducted into both the College and the Pro Football Halls of Fame. After playing in the NFL for nineteen seasons, Dawson had a successful career for many years as a sports broadcaster. The book that Dawson sent the Armstrong's was likely Len Dawson: Pressure Quarterback *(1970), by Len Dawson and Lou Sahadi.*

Bob DeMoss (1927–2017) was head football coach at Purdue University from 1970 to 1972, compiling an overall record of thirteen wins and eighteen losses.

GEORGE H. W. BUSH

August 24, 1971

Mr. Neil A. Armstrong
National Aeronautics and Space Administration
Washington, D.C. 20546

Dear Neil:

On September 9 we are having a reception at the Embassy—42A Waldorf Towers—for the Advisory Committee on the Environment, chaired by Senator Howard Baker. This is the citizens group which is supportive of the U.S. efforts involving the Stockholm Conference.

I am wondering if you would be willing to come to New York for this reception. Barbara and I would love to have you and Jan stay with us at the Waldorf that night. Afterward Howard Baker is having a small dinner which we would attend with you.

I had in mind your making a few comments at the reception about NASA's efforts through its earth satellites, stressing the great benefits to the environment that can come from NASA's program. A few slides would be in order, a movie, or just plain, raw Neil Armstrong at his best. This will be a very distinguished group of environmentalists on the Advisory Committee. In addition, we are inviting some of the top people from the U.N. and some private individuals in New York who are active in regard to the Stockholm Conference. I sure hope you can do it. I think it would be of advantage to NASA, but putting all that aside, it would sure be fun for the Bushes to see the Armstrongs again.

Warmest personal regards,

George
George Bush
The Representative of the United States of America
to the United Nations

Future forty-first president of the United States George Herbert Walker Bush (1924–2018) became America's ambassador to the United Nations in 1971, appointed by President Nixon with the unanimous support of the Senate; he served in the position into 1973. Armstrong would have many contacts with George Bush in the ensuing years, especially in Bush's roles as chair of the Republican National Committee (1973–1974), U.S. vice president, under Ronald Reagan (1981–1988), and as U.S. president (1989–1993). On July 20, 1989, the twentieth anniversary of the Apollo 11 moon landing, President George H. W. Bush stood on the steps of the National Air and Space Museum in Washington, D.C., and, backed by Armstrong and his Apollo 11 crewmates, announced his new Space Exploration Initiative (SEI), which he hoped would put America on a track to return to the Moon and eventually push on to Mars. Largely panned as a bad idea by the public, the media, and most in Congress immediately following its announcement, Bush's SEI proposal nonetheless came at a key turning point in NASA's history and ultimately contributed to the success of the International Space Station.

Senator Howard Baker was the Republican senator from Tennessee, Howard Henry Baker Jr. (1925–2014), who came to serve as Senate minority leader, then Senate majority leader. Known as the "Great Conciliator," Baker was highly regarded as one of the most successful senators for brokering compromises, enacting legislation, and maintaining civility—for example, his lead role in the fashioning and passing of the Clean Air Act of 1970 with Democratic senator Edmund Muskie of Maine. Baker is best remembered, though, for his influential role as the ranking minority member of the Senate Watergate Committee, which investigated the Nixon Watergate scandal. Baker famously asked aloud, "What did the President know and when did he know it?"

The work of the Advisory Committee on the Environment, chaired by Senator Baker, was preparatory to U.S. participation in the United Nations Conference on the Human Environment, convened in Stockholm, Sweden, in June 1972. With a growing interest in conservation issues worldwide, the idea behind the "Stockholm Conference" was to lay a foundation for global environmental governance. The conference's final declaration was a forceful statement of the finite nature of Earth's resources and the necessity for humanity to safeguard them. The Stockholm Conference led by the end of 1972 to the creation of the United Nations Environment Programme (UNEP), whose purpose was to coordinate global efforts to promote sustainability and safeguard the natural environment. Significantly, though the 1972 conference was attended by delegations from 114 governments, it was boycotted by Soviet bloc countries because of the exclusion of the German Democratic Republic (East Germany), which did not hold a U.N. seat at the time.

"Unfortunately, I will be in Norway"

September 1, 1971

Honorable George Bush
United States Representative to the United Nations
799 United Nations Plaza
New York, N.Y. 10017

Dear George:

Thank you for your kind invitation to join you at the reception in honor of the Advisory Committee on the Environment. Janet and I certainly would like to be able to join you for this occasion. Unfortunately, I will be in Norway attending a meeting of NATO's Advisory Group for Aerospace Research and Development, to which I am a National Delegate. I must apologize for making myself available and then reneging on our first invitation.

I will be leaving NASA at the end of September to accept a Chair at the University of Cincinnati. I am looking forward to this idea of sharing what I have learned from my experiences with the younger generation. Of course, I'll enjoy the freedom of some research work in areas of interest that I have been postponing for a decade or more.

I am still available to help wherever and however I can.

I have ordered a photograph for Ambassador Phillips and will forward it with pleasure as soon as it is available.

Janet joins me in sending her best personal regards to all the Bushes.

Sincerely,

Neil

Neil A. Armstrong
Deputy Associate Administrator for Aeronautics
Office of Advanced Research and Technology
NASA Headquarters
Washington, D.C. 20546

Ambassador Phillips is Christopher Hallowell Phillips (1920–2008). In 1970, President Nixon nominated Phillips to be the "permanent representative" of the United States to the United Nations Security Council, with the rank of ambassador.

MARGE SCHOTT

October 8, 1973

Mrs. Charles J. Schott
c/o Charles J. Schott, Inc.
300 American Building
Cincinnati, Ohio 45202

Dear Marge:

I was away from the university for the summer and out of the country in September so I didn't get your letter (via phone) till immediately prior to the dinner you hosted. I'm certain this was an inconvenience to you but I accept no responsibility.

I didn't understand when someone told me they had seen I was a member of your "Committee of 100." Now that I've read the "negative response indicates approval" paragraph in your letter I can see how that all happened. It is, however, a marginal technique for conducting business.

CELEBRITIES, STARS, AND NOTABLES

Enclosed is my contribution. I hope "Zoo 100" was a resounding success and admire your initiatives in making the whole thing possible.

Sincerely,

Neil A. Armstrong
Professor of Aerospace Engineering
University of Cincinnati

Mrs. Charles J. Schott—better known as Marge—was the majority owner of the Cincinnati Reds professional baseball team from 1984 to 1999. She was a colorful and controversial personality. While Reds owner, she was famous for bringing her dogs, Schottzie and later Schottzie 02, onto the field at Riverfront Stadium. In 1999 she was forced to sell her controlling interest in the Reds due to slurs she had made toward African Americans, Jews, and persons of Japanese ancestry, but she managed to retain a minority ownership in the team. Marge Schott was the third woman to own a North American major league team without inheriting it and the second woman to buy an existing team rather than inheriting it; she was also the first woman awarded a General Motors dealership in a major metropolitan area. She contributed to many Cincinnati area charities, including the Cincinnati Children's Hospital and Cincinnati Zoo. Armstrong occasionally attended Reds games with his family and on one opening day in Cincinnati made the ceremonial pitch into home plate.

"I am returning your check"

December 12, 1973

Dear Neil,

Thanks for your letter and check for Zoo 100. As you can see, I'm a little behind with my social & civic stuff. Sorry there was a mix-up on the whole thing with your being out of town.

 I am returning your check and you can save the 100 bucks in case I ever get us all involved in another charity or civic venture!

 I really wish you and Janet could have joined us—it really was fun, what with elephants, chimps, tigers, etc., running around!

 Thanks again, Neil, I appreciate your interest.

Sincerely,

Marge

P.S. Thought you might enjoy these Zoo 100 mementos. Also am enclosing an invite for the 26th to a normal party. I hope you and Janet will join us.

P.S.S. I'm also studying up on the art of football in case I ever sit next to your son Mark again!!

DR. CARL SAGAN

May 27, 1980

Dr. Carl Sagan
Center for Radiophysics and Space Research
Space Sciences Building
Cornell University
Ithaca, New York 14853

Dear Carl:

Your call was appreciated and your assistant, Shirley Arden, has forwarded the draft material, which clarifies your intent to a large degree.

It is an interesting concept. It may well be that the organization can provide a function not possible with the existing technical society structure and popular-level organizations. Certainly your leadership can provide a running start and the Advisory Board lends a degree of credibility and class.

While existing mechanisms have not been effective toward the stated objectives, it may be that further fragmentation could be counterproductive rather than synergistic. I know that you have given the matter considerable thought before launching the program.

As for me, I will wait. My existing commitments preclude accepting additional responsibilities, and if the organization is to succeed, it will require more than tokenism on the Board.

Should the organization become a reality, I will certainly support it with a membership. Thanks for the honor of the invitation.

All good wishes.

Sincerely,

Neil A. Armstrong

In 1980 Sagan (1934–1996) worked to create an organization known as the Planetary Society with cofounders Bruce R. Murray (1931–2013), a planetary scientist, and Louis B. Friedman (b. 1941), an astronautical engineer and space spokesperson. The mission of the Planetary Society is to "empower the world's citizens to advance space science and exploration." Its mission statement can be found at https://www.planetary.org/about/.

"Personally do not expect to participate in any substantial manner"

March 20, 1994

Prof. Carl Sagan
11 Tyler Road
Ithaca, NY 14850

Dear Carl:

Thank you for your thoughtful letter and suggestions regarding the Apollo 11 25th anniversary. There is apparently a substantial level of interest in the event as evidenced by the large number of planned television specials and other commemorative activities. I personally do not expect to participate in any substantial manner in those events. Mike and Buzz will, of course, make their own decisions.

You will recall that President Bush provided a ringing endorsement of planetary exploration and manned space flight on the Apollo 11 20th anniversary. That initiative was less than successful when the Congress did not support it in the budget.

We will be considering all the opportunities that are brought to our attention and undoubtedly will get involved in those which we can unanimously support.

Thanks again for your thoughts.

Sincerely,

Neil A. Armstrong

cc: Mike & Buzz

A RELUCTANT ICON: LETTERS TO NEIL ARMSTRONG

DR. HENRY HEIMLICH

March 16, 1982

Dr. Neil Armstrong
Route 2
Lebanon, OH 45036

Dear Neil:

I am planning visits to Beckley, West Virginia, in April, May, and June. Several patients will have the transtracheal tube inserted and it will be possible to see in follow-up those patients who have had the procedure done in the past. I think you would be most gratified to see the clinical results of our research.

Please send a few dates that would be satisfactory. I think June would be best for you hitting the little white ball. Please advise.

Best to all.

Sincerely,

Hank
Henry J. Heimlich, M.D.
Professor of Advanced Clinical Sciences
Xavier University
Cincinnati, OH 45207

While teaching at the University of Cincinnati in the 1970s, Armstrong joined a small team of researchers in the new Institute of Engineering and Medicine. The group included, besides Armstrong, George Rieveschl, a UC chemist famous for his invention of diphenhydramine (Benadryl), the first antihistamine; Edward Patrick, an electrical engineering professor; and Dr. Henry Heimlich, Cincinnati's famous inventor of the Heimlich maneuver. In his March 1982 letter to Armstrong, Heimlich was referring to a series of clinical trials he was conducting in West Virginia involving what he called "respiratory rehabilitation with a transtracheal oxygen system." Heimlich chose West Virginia for his trial because what he was researching was a more effective treatment for rehabilitating chronic obstructive pulmonary disease (COPD) patients, many of which were coal miners. His procedure involved administering oxygen continuously through an intravenous catheter inserted transtracheally. According to a paper Heimlich (1920–2016) published about his concept in the November 1982 issue of Annals of Otology, Rhinology and Laryngology, *his patients experienced an immediate sensation of being able to breathe more easily and began ambulating the day of the procedure, with improved nutrition and a return to many normal activities.*

GEORGE "SPANKY" MCFARLAND

May 26, 1985

Dear Neil:

Just a short note to ask if you received an invitation to play in my Celebrity Golf Classic Sept. 28 & 29? I mailed it in April and knowing the Postal system, it may have gone astray. If you didn't get it, please advise and I will send another.

    Fred MacMurray, Tom Poston, George Gobel, Arte Johnson, Phil Harris, George Kirby, and Andy Williams (if he's not working) have agreed to come. Bob Hope is still a "maybe," but I feel he will come. I would really like to have you on the list; your presence would certainly add a real touch of class and I know you would have a good time. Let me know.

Best regards,

Spanky
Spanky McFarland
Keller, Texas

George "Spanky" McFarland (1928–1993) gained fame as a child actor in the Our Gang *series of short-subject comedies of the 1930s and 1940s. The* Our Gang *shorts were later syndicated to television as* The Little Rascals. *As an adult, he was indelibly typecast in the public's mind as Spanky from* Our Gang, *and found little opportunity to continue his acting career. At age twenty-four in 1952, McFarland joined the U.S. Air Force. After leaving the military in the mid-1950s, he hosted an afternoon children's show on television in Tulsa, Oklahoma. When that show ended, he took different odd jobs: selling wine, operating a restaurant and nightclub, and selling appliances, electronics, and furniture. He made occasional personal appearances on television talk shows and in the mid-1980s helped launch The Nostalgia Channel on cable television. Spanky lent his name and celebrity to help raise money for a number of charities, primarily by playing in golf tournaments. His own namesake charity golf classic for sixteen years, held in Marion, Indiana, was the event to which he was inviting Neil. It is not known whether Neil attended, though he loved golf and did play in a number of charity events.*

DR. GERARD O'NEILL

July 2, 1985

Dr. Gerard K. O'Neill
President & CEO
Geostar Corporation
101 Carnegie Center, Suite 302
Princeton, NK 08540

Dear Gerry:

Thank you for giving me the opportunity to comment on your forthcoming article for "Discover."

First let me say that I think the concluding paragraph is superb. On the other hand, I'm not certain that the opening paragraph is necessary. It might seduce the careless reader into inferring that the opinions expressed are those of the Commission.

Regarding the middle part (between the paragraphs mentioned above), I have no objections regarding questions of fact. As you would expect, many of your conclusions from the available facts would not mirror my own. The opportunity is certainly available, however, for others to public their own views.

In any case, you have my congratulations for an interesting and provocative article.

See you in Houston.

Sincerely,

Neil A. Armstrong

In this letter Armstrong is reacting to an article that Gerard K. O'Neill (1927–1992) has drafted for publication in Discover *magazine that summarizes the findings of the Thomas O. Paine–led National Commission on Space. It would be interesting to know exactly what conclusions from the available facts did not mirror Neil's own conclusions.*

DR. HERMANN OBERTH

May 8, 1989

Prof. Hermann Oberth
c/o Arno Breker Society International, Inc.
P.O. Box 279
Clarence, NY 14031

Dear Prof. Oberth:

I am delighted to be able to add my congratulations on the occasion of your 95th birthday.

You have breathed life into the age of space, nurtured it, and watched it grow to a productive member of the human existence. Few will have witnessed so much and so personally in any field of endeavor.

Please know that, on this occasion, you have the congratulations, affection, and warmest of wishes for the years ahead.

Sincerely,

Neil A. Armstrong

Armstrong knew very well that Hermann Oberth (1894–1989), an Austro-Hungarian-born German physicist and engineer, was one of the founding fathers of rocketry and astronautics. In 1923 Oberth published the book Die Rakete zu den Planetenräumen (The Rocket into Interplanetary Space), *which gained him widespread recognition. He launched his first rocket on May 7, 1931, at a site near Berlin and in the ensuing years became a mentor to the young Wernher von Braun, working together with him in Germany through the end of World War II and after 1955 in the United States (though he returned to Germany three years later).*

A RELUCTANT ICON: LETTERS TO NEIL ARMSTRONG

Neil may have met Oberth more than once, but those details are unknown. Oberth did return to the United States in July 1969 as a special invited guest to witness the launch of Apollo 11 from Kennedy Space Center.

It is interesting that Armstrong mailed his letter to Oberth through the Arno Breker Society International, Inc. Arno Breker (1900–1991) was a German architect and sculptor who was greatly honored by Nazi Germany, with his art heartily endorsed by Hitler's regime as the antithesis of "degenerate" art. In the decades after the war, Breker produced a number of outstanding works and there were several efforts to rehabilitate his reputation, usually producing backlashes from anti-Nazi activists, including controversy in Paris when some of his works were exhibited at the Centre Georges Pompidou in 1981. As had been the case with Oberth, von Braun, and a number of the German rocketeers who worked for the German Army during World War II, Breker's admirers argued that despite being a member of the Nazi Party, he had never been a supporter of Nazi ideology but had simply accepted its patronage. How and why Armstrong knew how to reach Oberth in 1989 through the Arno Breker Society is unknown. Oberth died in Nuremburg, West Germany, on December 29, 1989, some six months after Neil sent him the ninety-fifth-birthday wish

DRS. HAROLD KLEINERT AND JOSEPH KUTZ

September 21, 1994

Dr. Harold E. Kleinert
Dr. Joseph E. Kutz
Kleinert, Kutz & Assoc.
Hand Care Center
Suite 700
225 Abraham Flexner Way
Louisville, Kentucky 40202

Dear Drs. K & K:

About midway through your partnership, a man with a severed finger, the result of a farm accident, was referred to your office. The man was flown down to Louisville from Ohio, and you were willing and able to begin operating within 7 or 8 hours of his injury. You missed cocktails and dinner that evening because of this Buckeye's poor choice of timing. Dr. Tsai also was called in so he could miss cocktails and dinner.

Well, many years have elapsed, and the fellow's finger is still attached. I would like to tell you that he plays the piano like Eddie Duchin, but that would be stretching the truth more than good taste will allow. He is able to hang on to the golf club with a full swing, and is able to use that finger—and nine others—well enough to type this letter.

So, on this special evening, I send my congratulations and thanks, with my very best wishes for the years ahead.

That Ohio Farmer,

N
Neil A. Armstrong

In November 1978, at his farm outside of Lebanon, Ohio, Neil severed the ring finger of his left hand when his wedding ring caught on a door while jumping off the back of a truck. He was rushed by air to an expert team of microsurgeons at the Jewish Hospital in Louisville, Kentucky, where a successful emergency surgery was performed (Neil's finger became fully functional except the uppermost joint). Drs. Kleinert and Kutz were his surgeons.

SIR EDMUND HILLARY

July 20, 1995

Mr. Neil Armstrong
c/o Ms. Vivian White
77 Columbus Ave. E.
Lebanon, Ohio 45036
U.S.A.

Dear Neil:

Just a short note to thank you and to bring you up to date regarding the print signing project you participated in for Special Olympics Nepal.

First of all, let me say it was your kindness in agreeing to autograph the prints of "Victory" that helped us raise the major portion of funds that allowed us to bring the Special Olympic team over from Kathmandu, Nepal.

The team of fourteen mentally challenged Nepali's who were able to participate in what will seem to them the event of a lifetime. The World Games took place from July 1 to July 10, 1995 at Yale University in New Haven, Connecticut, U.S.A.

Your prints of "Victory" by Lorne Winters were forwarded some time ago.

Thank you once again for your generous humanitarian effort in support of this worthy cause.

Sincerely,

Ed
S. P. Hillary
Special Olympics Nepal

In April 1985 Armstrong made an expedition to the North Pole under the direction of the professional expedition leader and adventurer, California's Michael Chalmer Dunn, and in the company of the world-famous climber of Mt. Everest, Sir Edmund Hillary, Hillary's son, Peter, and Pat Morrow, the first Canadian to reach Everest's summit. Neil maintained an occasional correspondence with Edmund Hillary until his death at age eighty-eight in January 2008. Neil also corresponded with Tenzing Norgay, the Nepalese Sherpa mountaineer who had guided Hillary to the 29,029-foot summit of Mount Everest on May 29, 1953.

JONATHAN WINTERS

April 13, 1997

Montecito, Calif.

Dear Neil and Carol—

Again I want to take this opportunity to tell what a fantastic evening we both had at "The Lone Sailor" dinner this past Friday evening. I've been to a number of dinners over the years but this was to top them all. Again Neil, I can't thank you enough for making that long trek from Cincinnati. And along with Eileen, Carol you were the prettiest lady there.

I'm enclosing a small but important piece of American naval history—a picture of Rear Admiral Farragut. I don't know what ship he was on but rest assured this is his signature. I've had it for a number of years in my collection and I wanted you to have it. I do hope we'll be able to cross paths again before too many months go by. And please if you're ever out this way, don't hesitate to pick up a phone and call us. It's [phone number withheld].

With deepest affection to you both.

Jonathan
Jonathan Winters
Montecito, Calif.

Fellow Buckeye Jonathan Winters (1925–2013) was one of Armstrong's favorite comedic talents. His range of comic characters, talent for mimicry, and brilliance at eccentric improvisation vaulted Winters to stardom in the 1950s and 1960s. He had three popular television comedy series: The Jonathan Winters Show *on NBC in 1956, a variety show by the same name on CBS from 1967 to 1969, and the syndicated* The Wacky World of Jonathan Winters *from 1972 to 1974. Neil did his best to watch as many of Winters's shows as he could. Winters was perhaps best known for his appearance on* The Tonight Show with Johnny Carson, *often causing Carson to double over with laughter. Winters appeared in a number of movies, often in cameos as eccentrics. He reached his widest movie moments with his wacky performance as the van driver in 1963's* It's a Mad Mad Mad Mad World. *He was also a regular on the game show* Hollywood Squares. *A talented voice-over artist, he played Papa Smurf on* The Smurfs, *an animated television show. Winters received the Kennedy Center's second Mark Twain honor for humor in 2000. His best-known act was his portrayal of a character he invented—Maude Frickert—a plump old woman with round glasses and a perverse sense of propriety. Winters greatly influenced comedians Dana Carvey and Robin Williams, both of whom considered him their mentor, with Winters making several appearances on Williams's 1978–1982 ABC sitcom* Mork & Mindy. *He had a serious side and suffered several bouts of depression in is life. As a young man, he studied at the Dayton Art Institute, where he developed a flair for drawing and became an accomplished cartoonist and painter.*

In 1997 Winters was one of the recipients of the Lone Sailor Award, presented to him at the annual Lone Sailor Dinner; it was to thank Armstrong for attending this event, at his invitation, that Winters wrote to him in April 1997. This award is given to "Sea Service veterans who have excelled with distinction in their respective careers during or after their service."[40] *At the 1997 ceremony* the award recipients included not only Winters but also actor Ernest Borgnine (1917–2012), U.S. Senator Robert Kerrey from Nebraska (b. 1943), publisher Austin H. Kiplinger (1918–2015), California businessman and philanthropist Edmund W. Littlefield (1914–2001), and entrepreneur and co-owner of the Oakland Raiders football team Henry F. Trione (1920–2015). One can be sure that Neil personally congratulated each one of them that night.

The Lone Sailor Award got its name in association with the effort back in the 1980s to raise money for building a United States Navy Memorial in Washington, D.C., its centerpiece being a bronze sculpture of The Lone Sailor, unveiled in 1987 as a tribute to all the personnel of the U.S. sea services.

It is curious that Armstrong never received the Lone Sailor Award.

CLINT EASTWOOD

June 24, 2005

Mr. Clint Eastwood
Tehama Golf Club
25000 Via Malpaso
Carmel, CA 93923

Dear Clint:

Carol and I enjoyed your elegant Carmel Valley hospitality at Tehama. You have built a great golf course and a clubhouse that sits well on the crest. I'm certain the Pacific sunset is spectacular on a clear evening but Mother Nature managed to keep that hidden from us this time.

As you might imagine, I have been approached by a number of authors about writing my biography or autobiography. I picked Jim Hansen above many that are more widely known. After talking with him and reading some of his work, I became convinced that he could do the job better than most because of his diligence, his knowledge of technology, and his attention to detail. For me, he is the right man.

We didn't talk much about your interest in doing a film of the book. That will be a matter, primarily, between you and Jim. A film is not at the top of my priority list, but if it is made, I would hope that it would be a credit to all those associated with the project.

Thanks again for a very pleasant day in Carmel. If you ever find yourself in the Cincinnati area, I can take you to a quiet old Seth Raynor course that I am confident you would enjoy.

Sincerely,

Neil A. Armstrong

When Armstrong gave the green light to write his biography, even that agreement made news, because Neil had never agreed to work with any writer on his life story or given him access to considerable private interview time and his private papers. As soon as that news hit the papers, Hollywood came to talk to me and I had to have an agent for the first time in my career. Ultimately, the film rights to the book were optioned to Warner Brothers and to Clint Eastwood. Clint wanted to direct the film through his Malpaso Productions, and he invited Neil and me and our wives to his private golf club above Carmel Bay along the central coast of California to play a round of golf and talk about the possibilities. After First Man *came out in 2005, the option ran out with Warner Brothers, and, because of the highly significant challenges of turning the Armstrong story into a film, Eastwood and Warner Brothers decided not to renew the option. Universal Studios in league with Temple Hill Entertainment picked up the movie rights, eventually moving forward with director Damien Chazelle to turn* First Man *into a major motion picture. The film premiered in the fall of 2018 and to date has won a number of major awards, including an Oscar and a Golden Globe.*

TOM HANKS

June 29, 2005

Mr. Tom Hanks
Playtone
Santa Monica, CA

Dear Mr. Hanks,

Thank you most sincerely for your kind letter of June 7, 2005 describing your forthcoming IMAX production, "Magnificent Desolation; Walking on the Moon." I am delighted to learn that a man of your talent and experience is associated with and, I am certain, leading and influencing the project. The fact that you are working with Eric Jones further increases my confidence in the quality of the product.

You are most kind and thoughtful to offer an honorarium. However, as I did when you made a similar offer with respect to your Apollo television series, I will decline. Again, and with some regret, I admit, I did not assist you in your efforts and have earned nothing to merit such compensation. If you feel compelled to distribute the funds, I suggest you send them to the Astronaut Scholarship Foundation.

I do hope that I will be able to accept your kind invitation to attend the premiere, but my calendar currently has me scheduled to be in Europe at that time. Perhaps I will get lucky and that schedule will change.

I enclose my very best wishes and remain,
Sincerely

Neil Armstrong

Neil was also offered a chance to consult on the Tom Hanks/Ron Howard film Apollo 13 *(1995), but he declined that opportunity as well.*

Eric M. Jones (b. 1944) is an astronomer who worked until 1999 at Los Alamos National Laboratory on a variety of defense-related programs. At that time Jones began a serious study of history, with a visit to Houston provoking a strong interest in the Air-to-Ground transcripts that had been archived from the Apollo missions. This interest led him to start a major project to create an online Apollo journal, which annotated in great detail everything mentioned in the transcripts. The result is one of the richest, most factual sources on the history of the Apollo program: the "Apollo Lunar Surface Journal" and the follow-on "Lunar Flight Journal." Originally from Wantagh, New York, Jones currently resides in Albury, New South Wales, Australia.

DAVID L. WOLPER

April 4, 2008

Mr. Neil Armstrong
777 Columbus Avenue, Suite E
Lebanon, Ohio 45036

Dear Neil,

I read in the entertainment trade papers that they're going to make a movie about your life. I couldn't be more excited. I think the public would like to know more about you and this is a beautiful way of telling them.

I don't know if you're going to be coming out here, but if there is any help you want regarding the film or regarding the deal they make, etc., any information you need, you can use my brain. I've been in the business for fifty years and I know every twist and turn, so I offer it to you.

In any event, good luck on the film. Hope to hear from you soon, and hope to see you.

Best personal regards,

David
David L. Wolper
The David L. Wolper Company, Inc.
Hollywood, California

P.S. Here's the story from Variety, in case you haven't seen it.

BARRON HILTON

August 18, 2010

Mr. Neil Armstrong
[Street address withheld]
Cincinnati, OH 45243

Dear Neil,

Sending my belated but most sincere birthday wishes to you, along with the enclosed photo album of our trip to Alaska. I really enjoyed having you aboard the Silverado and look forward to many more trips with you and our great group of friends.

I'm sorry I couldn't be with you and your many friends and family to honor you on your 80th Birthday celebration on August 5, but hope you know I was with you in spirit. I happened to be in Alaska for my fifth trip, and just recently concluded my fishing season this past Sunday.

Hope you enjoy the album, as well as the special book dedicated to our friend Paul Thayer, which you should have received a few weeks ago.

Looking forward to seeing you in Wyoming next month.

Best regards,

Barron
BARRON HILTON
Beverly Hills, California 90210

"In recognition of your inspiring lives"

March 2, 2011

Mr. Neil Armstrong
[Street address withheld]
Cincinnati, OH 45243

Capt. Eugene Cernan, USN (Ret.)
[Street address withheld]
Houston, TX 77024

Dear Neil and Gene,

I am pleased to inform you that the projects at the Purdue Libraries and National Flight Academy have passed muster with the Conrad N. Hilton Foundation.

Following the guidance of my son Steve and the staff of the Hilton Foundation, I am pleased to inform you that our board has approved a $2 million grant, payable over two years, to endow the archivist responsible for processing, preserving and sharing your personal papers in the Purdue Flight Archives in perpetuity. Our board has also approved a $3 million grant toward completion of the National Flight Academy at Pensacola NAS. It also will be paid over a two-year period, but only after the balance of funding for the final phase of the project is raised. We feel this challenge grant will energize the campaign to outfit the facility and complete the project.

You have both been an inspiration to me and everyone who has met you over the years. In recognition of your inspiring lives, I have personally donated half of the grant amount to the CNHF. You are great ambassadors for the space program, and tremendous role models for everyone who has ever dreamed big dreams.

Out staff will be in touch with Purdue and the Naval Aviation Museum Foundation regarding details on the grant process. I appreciate the work you have performed on their behalf.

Best regards,

Barron
BARRON HILTON
Beverly Hills, California 90210

"Your exceptionally generous gift"

March 7, 2011

Mr. Barron Hilton
[Street address withheld]
Beverly Hills, CA 90210

Dear Barron,

Your letter arrived almost simultaneously with an e-mail from Purdue University's Dean of Libraries, Jim Mullins, announcing your wonderful gift to Purdue. And your letter surpassed that surprise with the news that the Conrad N. Hilton Foundation and yourself had also made a major conditional gift to the National Flight Academy in Pensacola.

Your exceptionally generous gift specifying and making possible the professional archiving of the papers of Gene and me is an extraordinary act on your part and, for me, it is very much appreciated.

I know the Purdue grant will be very well used. The library has been working diligently for a number of years to transform itself into a world class facility. And they have succeeded. Those skills and facilities were very important to me in selecting a repository for my library and papers.

I am a new member of the Board of Advisors of the National Flight Academy. They have a brand new facility and noble goals. But they have yet to demonstrate their value. As an advisor, I intend to do all I can to make certain that they successfully meet those worthy goals.

Barron, you are a good man and I am proud to call you a friend. Thanks so much for all that you do.

Sincerely,

Neil
Neil A. Armstrong

A RELUCTANT ICON: LETTERS TO NEIL ARMSTRONG

*"Preserving the personal papers and artifacts of
two of our most beloved Boilermakers"*

March 11, 2011

Mr. W. Barron Hilton
[Street address withheld]
Beverly Hills, CA 90210

Dear Mr. Hilton,

On behalf of Purdue University, thank you for the generous $2 million grant from the Conrad N. Hilton Foundation (CNHF) benefitting Purdue Libraries and ultimately the endowment of a permanent archivist. Your personal contribution of $1 million to the CNHF for this grant is a testament to your commitment to preserving the personal papers and artifacts of two of our most beloved Boilermakers—your friends and fellow Navy veterans—Gene Cernan and Neil Armstrong.

Your gift is a realization of Gene's dream to have his and Neil's legacies preserved at their alma mater. A permanent archivist will be devoted to developing the collection on human space flight, processing papers, building displays, digitizing collections, public speaking and developing online curriculum for the classroom. First priority will be the Cernan and Armstrong collections and working with these two American heroes to document their stories. I share your passion for flight and your belief that access to their historic journeys will continue to inspire people everywhere.

Gene and Neil represent well the "victories and heroes" of Purdue and are considered two of the most legendary members of our University family. Thank you, Mr. Hilton, for your trust and confidence in Purdue Libraries.

Sincerely,

France A. Córdova
President
Purdue University
Hovde Hall, Room 200
610 Purdue Mall
West Lafayette, IN 47907

CELEBRITIES, STARS, AND NOTABLES

cc: Gene Cernan
     Neil Armstrong

"244 pound tuna cow"

March 22, 2011

Mr. Neil A. Armstrong
[Street address withheld]
Cincinnati, OH 45243

Dear Neil:

Thanks for your recent letter expressing appreciation for the Hilton Foundation's grants to Purdue University and the National Flight Academy in Pensacola. We were both pleased and proud to support these causes on behalf of you and Gene.

    I thought you would like to see a photograph of a 244 pound tuna cow I caught on my recent trip to Socorro, Mexico.

    Look forward to seeing you for our fishing trip in Alaska July 19–23, with the same group as last year.

Best regards,

Barron
BARRON HILTON
Beverly Hills, California 90210

"Their historic journeys can continue to inspire people everywhere"

March 22, 2011

Ms. France A. Córdova, President
Purdue University
610 Purdue Mall
West Lafayette, IN 47907-2040

Dear Ms. Córdova:

Thank you for your kind letter of March 11, acknowledging the $2 million grant from the Conrad N. Hilton Foundation to benefit Purdue

Libraries and the endowment of a permanent archivist.

We were pleased and proud to assist in this regard to preserve Gene Cernan's and Neil Armstrong's collections on human space flight so that their historic journeys can continue to inspire people everywhere.

Sincerely,

Barron Hilton
BARRON HILTON
Beverly Hills, California 90210

cc: James L. Mullins, Ph.D.
    Gene Cernan
    Neil Armstrong

"You've made Neil and Gene feel better about it, too"

May 24, 2012

Mr. Elon Musk
Chief Executive Officer
Space Exploration Technologies
1 Rocket Road
Hawthorne, CA 90250

Dear Elon,

As you might recall, I picked up my FAI Gold Air Medal the same evening that you received your Gold Space Medal at the NAA dinner in Washington, D.C. I regret that we never got the chance to talk during the program that night.

I am writing to congratulate you and your Space X team on the successful launch of the Falcon 9 on its first supply mission to the International Space Station. I remember a fishing trip last summer when Neil Armstrong and Gene Cernan were preoccupied by their upcoming statement to Congress regarding NASA's request for funding for the manned space program. They felt NASA should have articulated a goal that would serve to inspire the nation.

I must say that you and Space X have done more to inspire Americans in a few short months than NASA has in the years since it decided to end the Shuttle program. I am heartened that private enterprise has managed to take the initiative, and offer some hope for the future of America's manned space program.

I have a feeling you've made Neil and Gene feel better about it, too.

Best regards,

Barron
BARRON HILTON
Beverly Hills, California 90210

7
LETTERS FROM A GRIEVING WORLD[41]

August 25, 2012

We are heartbroken to share the news that Neil Armstrong has passed away following complications resulting from cardiovascular procedures. Neil was our loving husband, father, grandfather, brother and friend. Neil Armstrong was also a reluctant American hero who always believed he was just doing his job. He served his nation proudly, as a navy fighter pilot, test pilot, and astronaut. He also found success back home in his native Ohio in business and academia, and became a community leader in Cincinnati. He remained an advocate of aviation and exploration throughout his life and never lost his boyhood wonder of these pursuits. As much as Neil cherished his privacy, he always appreciated the expressions of good will from people around the world and from all walks of life. While we mourn the loss of a very good man, we also celebrate his remarkable life and hope that it serves as an example to young people around the world to work hard to make their dreams come true, to be willing to explore and push the limits, and to selflessly serve a cause greater than themselves. For those who may ask what they can do to honor Neil, we have a simple request. Honor his example of service, accomplishment and modesty, and the next time you walk outside on a clear night and see the moon smiling down at you, think of Neil Armstrong and give him a wink.

Sincerely,

The Armstrong Family[42]

On August 6, 2012, one day after his eighty-second birthday, Neil Armstrong mentioned to his wife, Carol, that he was having chest pain. Carol tried to get him to drive to the hospital, but he resisted because his symptoms seemed to be getting better with antacids. Carol called his cardiologist, nevertheless, who asked them to come to the hospital immediately, directing them to Fairfield Mercy Hospital, a 293-bed facility located in a northern suburb of Cincinnati in Butler County, Ohio. Once there, Neil told the cardiologist, who over recent years had become Neil's good friend, something he had not told Carol: that he had been unable to walk up the small hill in front of his house to the mailbox without being out of breath. The cardiologist ordered an immediate stress test.

Neil's stress test results raised serious concerns about cardiac ischemia, a restriction of blood supply to the heart muscle. It was pertinent also that Neil was experiencing a severely low heart rate. That same day he underwent coronary artery catheterization, which showed extreme narrowing of a minimum of three coronary arteries. The cardiologist determined that coronary bypass surgery needed to be performed immediately—the next day. Given the apparent urgency of his medical situation, there was no discussion about getting a second opinion, or considering moving to a different hospital, such as the Cleveland Clinic or The Jewish Hospital–Mercy Health in downtown Cincinnati.

A heart surgeon who had been recommended by Neil's cardiologist performed the surgery at Fairfield Mercy Hospital the morning of August 7. A temporary pacemaker was implanted, the type of pacemaker universally used in coronary artery bypass graft (CABG) surgery, which involved the placement of temporary epicardial pacing wires at the completion of the surgery, before the chest was closed. This pacemaker was placed in case Neil developed an abnormal heart rhythm after surgery, so it could be quickly corrected. The wires exited Neil's body through the surgical incision site—in this case, through the sternum.

After surgery, Neil's heart rate remained very low, with the temporary pacemaker continuing to pace his heart. At this point Neil's cardiologist concluded that a permanent pacemaker needed to be implanted. An electrophysiologist—a cardiologist who specializes in electrical issues with the heart—met with Neil on August 13 and confirmed that he needed the permanent device.

By August 14, the day the permanent pacemaker was to be placed, Neil was feeling and acting considerably better. Much of the mental confusion associated with the aftermath of bypass surgery had cleared up, and he was able to walk slowly through the hospital hallways. He still had a great deal of chest pain when he coughed, but he did not complain about it. When family members who had been visiting him after his surgery left the hospital on the fourteenth, they anticipated that Neil would be heading home on the fifteenth, about twenty-four hours after pacemaker placement, barring unforeseen circumstances. The permanent pacemaker was placed as scheduled on August 14 by the electrophysiologist. The procedure was uneventful. Neil told Carol he was hungry, and he joked with the nurses as they rolled him back to his cardiac ICU room.

Once the permanent pacemaker was placed, there was no longer a need for the temporary pacemaker and pacing wires. The wires were external to the skin but led through the incision site into the chest where they were implanted into the epicardial tissue. When these wires were pulled, at 1:15 p.m. in Neil's cardiac ICU room, they were supposed to slowly unwind, letting go of the heart tissue, then slide out through the skin. A dressing would then be placed over the small hole that was left. (Some doctors prefer not to pull the wires, but instead cut them at the level of the skin, in order to avoid the risk of heart injury during the pull.)

Neil remained comfortable after the wires were pulled and was eating lunch. Carol, who had slept in a chair at the hospital every night since Neil's admission, was across the hall from his room when at 1:35 p.m. a code blue was called, signifying an immediately life-threatening emergency, generally a cardiac or respiratory arrest. Quickly an echocardiogram—an ultrasound of the heart—was done, which showed blood pooling around the heart. With so much blood remaining within the pericardium, a fibrous sac surrounding the heart, such great pressure built up around the heart that it prevented its chambers from filling properly. (Cardiologists call this "tamponade.") When this occurs a person's blood pressure can drop to zero, meaning that blood would have stopped flowing through the patient's arteries, which can be quickly fatal. It is not known exactly how low Neil's blood pressure dropped, but it was to a very low level. Once this occurred, his brain did not receive adequate blood, and thus oxygen, to stay alive. The family was informed that there had been no cardiothoracic surgeon on call in the hospital, no specialist immediately available in case such a complication arose from the pacing wire extraction. Within a short amount of time Neil's brain suffered significant irreversible damage.

At 2:10 p.m. the cardiac catheterization lab attempted to remove the blood from around Neil's heart with little success and no improvement in blood pressure. He was given strong medications to raise his blood pressure, as well as a blood transfusion. Even with these measures, his blood pressure did not normalize. At this point, the surgeon who performed Neil's original surgery, who had been called back from another area hospital, arrived at Fairfield Mercy.

At 2:32 p.m. Neil arrived in the operating room. His chest was opened and his heart was observed to not be pumping at all. The electrical rhythm of his heart at that time was ventricular fibrillation, a disorganized electrical pattern that leads to quivering of the heart muscle without any effective pumping. Some 2.4 liters of blood were removed from Neil's heart in the OR. This represented almost half the total amount of blood that would be expected to be in his body. After removal of the blood, Neil's blood pressure normalized, but not for fifteen to twenty minutes.

Tragically, for approximately seventy-seven minutes Neil's brain had been receiving significantly insufficient oxygen. There is uncertainty, and likely individual variation, regarding how much time a human brain can live without receiving oxygen. In Neil's case, a dreadful amount of time passed, leading to severe brain injury. During the emergency surgery, aside from the blood that was removed, Neil's surgeon had not found any ongoing area of bleeding. His assumption was that it was the pulling of the epicardial pacing wires that had caused the original bleeding, but that the area may have clotted off before the surgery, perhaps helped by the low blood pressure. In retrospect, a number of questions would emerge about Neil's case—most notably, why there was no thoracic surgeon in the building when the pacing wires were removed? Arguably, Neil's emergency should have been handled with immediate opening of the chest in the operating room, or even in the ICU room, if necessary, rather than in the catheter lab, but this depended on a surgeon being available.

Depending on one's definition of death, Neil did not die for eleven more days. As is the case in most situations when a patient is not conscious, an endotracheal (breathing) tube was placed to make sure that Neil did not aspirate. He was placed on a ventilator. His permanent pacemaker supported the beating of his heart. He received medications to increase his blood pressure in the catheter lab, as well as blood transfusions. Subsequently, he was able to maintain a normal blood pressure without medications. The fundamental issue was brain injury.

After the bleed, Neil never displayed any purposeful neurologic function. On August 15 a CT scan of his head was performed: the findings were consistent with anoxic brain injury—a brain injury due to an abnormally low amount of oxygen in the brain tissue. A neurologist came to consult, and an electroencephalogram (EEG) was performed. Carol and other family members discussed Neil's condition with Neil's chief cardiologist, who informed them that there was no expectation of meaningful neurologic recovery. A second CT scan was also performed. Carol kept a vigil at Neil's bedside throughout the coming days. Other members of the family also spent agonizing hours waiting desperately at the hospital for some sign of improvement. But Neil never displayed any intentional responses of any kind.

Understandably in shock, members of the family met together several times to discuss the situation and pray for answers. At least one of the meetings was attended by Janet Armstrong, Neil's first wife. Neil's cardiologist and his surgeon came to one meeting to discuss the chances for Neil's neurologic recovery. At one point, the cardiologist presented the option of giving Neil a permanent feeding tube and tracheostomy for his ventilator and putting him in a nursing home. Pressed by the family for a clearer prognosis, the doctor confirmed what the neurologist had concluded: there was "no meaningful chance of recovery." Neil had a living will that clearly stated he did not want to be maintained in a vegetative state. To ignore his wishes, or defer indefinitely making a decision, would be to deny Neil his wishes and autonomy at a point when he no longer had any power to control his own fate.

After this family meeting with Neil's doctor, Carol, who was Neil's designated medical power of attorney, accepted—with the most heartbreaking reluctance—that life support should be withdrawn. Neil's youngest son, Mark, wanted more information, so a second neurosurgeon was brought in for an opinion on August 20. This doctor's assessment was in line with the information already given to the family—that the chance for any degree of eventual recovery was slight and that recovery would mean life in a nursing home with complete dependence.

Although Neil was in a coma, he registered extreme pain whenever his ventilator was cleared. Carol told the nurse, "No more." On August 21, Neil's breathing tube was removed. His code status was changed to Do Not Resuscitate (DNR). Comfort measures only were to be employed. The family expected that he would pass within a few hours, though the doctors and nurses explained that even without life support, some patients hold on for many days. Neil hung on for four days, until August 25.

A private funeral for family and close friends was held on Friday, August 31, at the Camargo Golf Club in Indian Hill, the Cincinnati suburb in which Neil and Carol had lived since their marriage in 1994 and the club of which they were long-time members. An estimated 200 people attended, including Neil's relatives and close friends (as well as this author). There was tight security to keep out the press and the uninvited, a navy ceremonial guard, and a bagpiper. Mike Collins and Buzz Aldrin were there, as was John Glenn and Jim Lovell, along with several other astronauts, space program officials, and aerospace notables. Giving eulogies were Ohio congressman Rob Portman, a family friend, and Charles Mecham, Neil's long-time friend and former head of Taft Broadcasting. Neil's two sons, Rick and Mark Armstrong, presented short talks about their dad, sharing personal anecdotes—and some of their dad's favorite jokes—lifting the spirits of the grieving. Carol's son, Andrew Knight, read from 1 Corinthians, and her granddaughter, Piper Van Wagenen, read the 23 Psalm. Metropolitan Opera's Jennifer Johnson Cano, a mezzo-soprano, sang the pop standard "September Song," a favorite of Neil's and a metaphor comparing a year to a person's life span, from birth to death. At the end of the ceremony, everyone walked out onto the ninth fairway to witness a flyover of F-18 fighter jets peeling away in the missing man formation.

So loved and admired was Neil Armstrong that many Americans, led by Ohio congressman Bill Johnson, called for President Obama to grant him a state funeral, a highly formal event steeped in tradition and usually only held for former presidents. (President Obama did direct all American flags to be brought to half-staff throughout the nation until sunset on Monday, August 27, as well as "all United States embassies, legations, consular offices, and other facilities abroad, including all military facilities and naval vessels and stations."[43]) The state funeral did not materialize, but a large public memorial service was held on Wednesday, September 13, at Washington National Cathedral in the lovely northwest quadrant of the nation's capital. A magnificent Gothic structure, the cathedral was an especially appropriate place to hold the Armstrong service as its Space Window depicts the Apollo 11 mission and holds a sliver of Moon rock amid its stained glass panes. Before an overflowing crowd, Mike Collins led the mourners in a prayer. Eulogizing Neil was his good friend Gene Cernan, the Apollo 17 mission commander and last man to walk on the Moon, and Charles Bolden, the NASA administrator. One of Neil's favorite contemporary singers, jazz contralto Diana Krall, sang "Fly Me to the

Moon." The Reverend Gina Gilland Campbell read a passage from the book of Matthew, and the Right Reverend Mariann Edgar Budde delivered a homily.[44] The following day, September 14, Armstrong's cremated remains were scattered in the Atlantic Ocean during a burial-at-sea ceremony aboard the USS *Philippine Sea*. Neil's wife Carol, sons Rick and Mark, sister June Hoffman, stepdaughter Molly Van Wagenen, and stepson Andrew Knight and his wife Cristina were on board for the service. A U.S. Navy firing squad fired volleys in Neil's honor, followed by a playing of taps. As Secretary of the Navy Raymond E. Mabus asserted in advance of the burial, "Neil Armstrong never wanted to be a living memorial, and yet to generations the world over his epic courage and quiet humility stands as the best of all examples."[45]

The detailed narrative of Neil's heart surgery and how he actually died, provided below, were not known to the public until the third week of July 2019. What little the public knew had come from that short statement the Armstrong family released announcing his death on Saturday, August 25, 2012. But, as the world was to learn, the full story of what had happened to Neil was much more complicated—and extraordinarily sad.

On July 23, 2019 (three days after the commemoration of the fiftieth anniversary of the Moon landing), the *New York Times* published on its front page an article by reporters Shane Scott and Sarah Kliff about a medical malpractice suit members of the Armstrong family had filed against Mercy Health–Fairfield Hospital in Ohio shortly after Neil's death.[46] A few days earlier the newspaper had received by mail, from an anonymous sender, ninety-three pages of documents related to Neil's hospitalization and a resulting legal case, including copies of reports by medical experts for both the Armstrong family and the hospital; as the *Times* reporters explained, some of these documents, though marked "filed under seal," were publicly available at the Cincinnati probate court's website, confirming that the documents received by the newspaper were authentic. In the envelope there was also an unsigned note stating that the sender hoped disclosing the information would save lives. In the following days, a number of other media outlets, including the *Cincinnati Enquirer*, announced that they too had received that package of documents, and stories about the private details associated with Neil Armstrong's death broke all over the world.[47]

The breaking story also reported that Fairfield Mercy, in 2014, while its administrators and lawyers defended the care the hospital gave Armstrong, had paid members of the Armstrong family, notably Neil's two sons, Rick and Mark Armstrong, $6 million in a private settlement, thereby avoiding a malpractice lawsuit and the devastatingly bad publicity that likely would have resulted from it. A condition of the settlement was that both the hospital and the Armstrong family (that was, those who signed a non-disclosure agreement) would keep the story of what had happened to Neil, and the settlement, a secret. As the *New York Times* coverage made clear, Armstrong's wife, Carol, was not a party to the settlement and took no money, as she felt that her husband would have been opposed to taking legal action and would not have thought that anyone should benefit financially from his death. However, Carol did also sign an NDA, fearing that the family might remove her as the executor to Neil's will if she did not sign it.

As sad as it had been for everyone in August 2012, learning the truth of Neil's death was horrible. His death was not just sad, but tragic; it may not have needed to happen. For a man who could have died on so many occasions during his extraordinary life in air and space—in combat over North Korea, when test-flying highly dangerous unproven airplanes at Edwards, in fiery launches atop powerful rockets known to blow up at Cape Canaveral, while spinning dizzily and nearly blacking out in an out-of-control spacecraft following rendezvous and docking in Earth orbit, managing to eject from a lunar landing training vehicle a fraction of a second before the perverse machine exploded at Houston's Ellington Air Force Base, flying long and running out of fuel in *Eagle* on the way down to a rocky and cratered landing site on the surface of the Moon, just to list the most outstanding incidents when death breathed down his neck—it was immensely sorrowful and ironic to know the actual details of Neil's death following his heart surgery.

Fortunately, the world did not hear about them until years after he was gone. That way, in the news of his death, and at his public memorial service and private funeral, the focus could be on what was overwhelmingly good and positive about his life, without the sensational media coverage that would have inevitably overwhelmed the news of his death at the time. As his Apollo 11 crewmate Mike Collins said at Neil's private funeral, "He was the best, and I will miss him terribly." "You'll never get a hero like Neil Armstrong," added Apollo 8 astronaut Bill Anders. "He's going to be hard to top." With tears welling in his eyes, his old Purdue buddy Gene Cernan remarked: "He was the embodiment of everything the nation is all about. There's nobody that I know of that could have accepted the challenge and responsibility that came with being that with more dignity than Neil Armstrong."[48]

Similarly, this second book of letters to Neil merits a positive ending that is in keeping with the highly respectful and admiring statements made about Neil in the immediate aftermath of his death, unspoiled by today's dreadful knowledge of what happened to him. In that spirit, this final chapter publishes a fascinating and heart-rending collection of letters of condolence written and sent to Carol Armstrong in the weeks and months following Neil's death in August 2012.

It was incredibly difficult for Carol to read those letters the first time, let alone reread them eight years later and contemplate sharing them with the world. When I initially asked her if she might share some of the letters for this book, she balked: understandably, Carol wasn't sure if she could bear to look at them again. Many months later, she changed her mind. She resolutely looked again—very painfully—through the several hundred letters of respect and sympathy she had received following Neil's death, choosing eighty of them for possible inclusion here. I close this book with the letters of those who gave their permission to include them here—letters which in different profound ways capture the essence of who Armstrong was, what he meant to the letter writer, and what he meant to the world.

"WE WERE TRANSFIXED"

July 25, 2012

Dear Carol:

Susan and I were saddened to learn of Neil's death. Please accept our deepest sympathy. We wish that we could offer just the right words to bring you comfort. Please know that we are praying for you and for Neil.

Susan and I treasured our friendship with you and Neil. The evenings that we spent together were warm, joyful occasions. I especially remember the night that Neil retired from the Milacron board of directors. Our meeting was held at Lake Como and we were served a magnificent dinner. It was a beautiful scene with the lake and mountains surrounding us. As we relaxed over one last glass of wine, Neil rose and reflected on his past. He spoke with a sense of wonderment over the events of his personal and professional life. As always, he gave credit to others for his achievements. We were transfixed by the amazing story that he told in modest, but poetic, words. It was a magic moment.

Neil and I had a very warm friendship. We often saw each other at civic gatherings and at the Commercial Club. We would greet each other, shake hands, and then he always gave me a hug. I cherished those hugs and wish that I could have one more.

Carol, we are thinking about you during this sad time. Please call us if there is any way we can be of help.

Sincerely,

Joe Pichler
Cincinnati, Ohio
Ret. Chairman & CEO The Kroger Co.

Joseph A. Pichler (b. 1939) served as chair of The Kroger Co. from 1990 to 2004 and as CEO from 1990 to 2003. He helped guide Kroger through a takeover battle and financial restructuring in 1988 and a decade later engineered the largest merger in Kroger's history. Under his leadership, Kroger grew from approximately 1,200 supermarkets and $20 billion in annual sales to more than 2,500 stores, $54 billion in annual sales, and 290,000 associates. On Pichler's watch Kroger made many acquisitions, including Fred Meyer, a West Coast chain of grocery and general merchandise stores. Born in St. Louis as the fifth of six children, Joe earned his undergraduate degree at the University of Notre Dame and master's and doctoral degrees in business from the University of Chicago. For fifteen years he taught at the University of Kansas, becoming the dean of its school of business. He then moved into the business world itself, accepting a management position with the Dillons Food Stores in central Kansas, from which he eventually moved on to the leadership position with Kroger.

Neil knew Joe Pichler through their mutual service on the board of directors of Milacron, Inc., a company based in Cincinnati that produces high-quality machine tools, plastics processing technologies, and industrial fluids for the automotive and aerospace industries. Reincorporated in 1970 from what had been the Cincinnati Milling Machine Company (founded in 1884), Milacron today has some 4,000 employees working in 150 facilities in North America, Europe, and Asia. Armstrong served on its board of directors from 1980 to 2000. Since 2001, Milacron has awarded an annual Neil A. Armstrong Award to an "exceptional career employee who demonstrates the pioneering spirit and professional achievement levels of the former astronaut." Neil and Carol also knew Joe and his wife, Susan, through the Pichlers' many civic and charitable activities in the Greater Cincinnati area.

"I WILL ALWAYS REMEMBER HIM"

August 25, 2012

Dear Armstrong Family,

Today I heard with great sorrow about the departure of Neil. One and a half years ago I walked with him hand-in-hand between endless rows of Space Camp students. Neil then presented me the Lifetime Achievement Award for Education. I am most fortunate to have known this great man. He has been one of the two greatest men I ever knew: Wernher von Braun who dreamed of traveling to the moon and Neil who went there where no man had ever been.

    I feel with you the great sadness of Neil's leaving. I will always remember him.

    I wish you peace and hope.

Georg von Tiesenhausen
Huntsville, AL

Dr. Georg von Tiesenhausen (1914–2018) was a distinguished member of Wernher von Braun's rocket team. His February 22, 2011, letter to Neil is included in chapter 4, "Fellow Astronauts and the World of Flight."

"THEY'RE FLYIN'"

August 25, 2012

Mrs. Armstrong,

My father, Dave Stephenson, knew your husband when they were Navy men at Purdue. In fact, I believe you met my dad some years ago at a small reunion of friends from those days.

    My dad had heart surgery in May, and he never quite recovered. He died on August 8th.

When I heard about your husband's death, I felt such sympathy for you. But there was more to it for those of us who knew that once these two had been daring young men. "Neil A. passed today," my brother texted to me. "They're flyin'."

In 1976 when Mr. Armstrong came to Pennsylvania to participate in Bicentennial festivities at Valley Forge Park, he spent an evening with my parents in their nearby home. "It was real nice," my dad told me later. "We just talked about old times. We didn't even mention the moon."

You are in my thoughts.

Anne Carpenter
Phoenix, AZ

Anne Carpenter's father was David S. Stephenson (1930–2012), one of Neil's freshmen classmates at Purdue. Like Neil, Stephenson attended Purdue on a U.S. Navy scholarship and was called to aviator flight training at Naval Air Station Pensacola after only three semesters at the university, becoming a part of the forty–naval cadet preflight training group designated as Class 5-49, the fifth class to begin training at NAS Pensacola in 1949. In approving this book's use of her condolence letter to Carol Armstrong, Ms. Carpenter wrote: "My family knew Neil years ago when he was stationed on the USS Essex. *His good friend, Ken Danneberg and [Ken's] bride, had rented our little cottage next to our house. Every Saturday morning Neil would come to have breakfast with Ken and Florence and I would look out the window and sigh. He was a very good-looking young man and I was a freshman in college. Neil and Ken spent many weekends working on an old sailboat and then they took up sailing, with amusing mishaps. They remained good friends for the rest of their lives and we often heard of Neil through Ken, who remained our good friend. Ken called him affectionately 'the kid' because of his boyish looks. About ten years ago, Ken asked us [my husband Pat and me] if we would host a party for the squadron in our garden. That was when we met Carol and I was so happy that my family could meet Neil and Carol. She was a very lovely addition to his life. That afternoon, my grandson Matthew asked Neil what he was like when he was young and Neil replied that he was very 'bookish.' Matthew then asked him for an autograph and he gave it, perhaps because Matthew was about 6 years old. Ken remarked, 'he never signs autographs!' Forgive me for sharing all this but I wanted you to know about the past history, too, and about Neil Armstrong's time in this little*

A RELUCTANT ICON: LETTERS TO NEIL ARMSTRONG

navy town of Coronado, where so many people happily return when careers are over. With gratitude and appreciation, Anne T. Carpenter."

Anne's husband, Pat (who died in 2018), also wrote a condolence card to Carol Armstrong, dated August 26, 2012. The following are excerpts from that letter:

"Neil's 'small step for man' was truly historic, the implications of which will only be fully understood by future generations. We are proud to say that we have been among the honored few that have had personal contact with this genuine hero and can vouch for his character. He really did 'make a difference' in our country and in our world. . . . We were told about the heart problem a few weeks back [by Ken Danneberg] and assured that recovery was progressing favorably. Along with so many others, we were shocked to hear the news of Neil's passing on the TV news. Since then we have been eagerly reading, looking, and listening to accounts and news coverage to fill in some of the blanks in our version of the 'Neil Armstrong Story.' The universal praise, admiration, and appreciation from all quarters have been so gratifying. We know that no words or gestures can ease the pain and grief that you are feeling but will pass on a quote attributed to 'Dr. Seuss' that went something like: 'Don't cry over what is. Smile over what was.' We know that, after your very necessary period of mourning you will be doing just that. We—and all the country . . . in fact, all of the world—will recognize, respect, and acknowledge with pleasure that the man you were married to, and who departed much too soon, Neil Armstrong, is now a revered man of history."

"IT IS VERY HARD, THIS NIGHT"

August 25, 2012

Dear Madam Armstrong and her family,

It is with one very big sadness that I learnt the death of your husband and my hero, Neil Alden Armstrong, and I am sincerely anxious to present you my most sincere condolences and I am anxious to share your punishment, you, your family and all the humanity have just lost a great man, for me, it is more than the hero who I have just lost, it is a member about my family, that I always have dreamt to meet. I have to love two men in my life: my Father and Neil Armstrong, two left me in the same circumstances. They had the same age. My father was born in June, 1930, Neil in August, 1930. My father is to die two days after his operation in November, 1991. I was six years old, the age of Neil's son

Mark, when my Father me accept to look at Neil walked on the moon and this day there, I always remember it. I said that I wanted to become Neil Armstrong. The hard reality of the life did not allow it me, either to become Neil Armstrong, or even to meet him. I was lucky to be able to correspond with him in the 90s, after that not for more possible. As I write these lines, I have tears in my eyes, there now. I lived in a world where Neil Armstrong is not there any more and it is very hard, this night. I have to look at the moon and think of Neil and wish him a safe journey among the stars.

To excuse I for being so long but for me Neil Armstrong was more than a simple hero and I want to join to your sorrow, to share your punishment because, as you, I have just lost a dear human being.

Sincerely Yours,

Michel Sergent
Bourg-Fidele, France

The identity of the Monsieur Michel Sergent that wrote this letter is unknown. The best-known Michel Sergent in France has been a member of the Senate of France, a mayor, and head of La Fédération départmentale de l'énergie du Pas-de-Calais, which places him geographically in the very far north of France, not far from Bourg-Fidèle; however, that man was born in 1943, and the writer of the condolence letter seems to have been born in 1963. Every possible effort was made to identify the correct Michel Sergent, unsuccessfully.

"HE WAS SO MUCH MORE"

August 25, 2012

Dear Mrs. Armstrong,

Like so many people across the country and around the world, I was deeply saddened to hear that your husband passed away today. As you might know, Neil and I exchanged many e-mails—and many thoughts during the past several years. I was pleased that he sought my comments and suggestions in his effort to help get America's space program back on course. And I was honored to be able to call him a friend. (Our mutual friend, Jack Schmitt, brought us together three years ago when President

Obama cancelled the Constellation Program.) Through our exchanges, I came to more fully appreciate Neil's wisdom—and his wit. He was a remarkable man. But I don't need to tell you that.

What I *do* need to tell you is how very sorry I am about your loss. To the world, Neil Armstrong was a genuine hero. But for you, he was so much more. I hope you can find some comfort in knowing that Neil is now in God's loving arms . . . and that your loss is shared by so many others.

I am so very grateful that I crossed paths with Neil—if only in cyberspace. And I am so very sad that you and I are now 'crossing paths' under these circumstances. But I did want you to know that you are in my thoughts and prayers, as are the rest of Neil's family and friends. Please know that I'll be winking at the Moon—and thinking of Neil every time I do. God bless.

Sincerely,

Bill
William F. Mellberg
Park Ridge, IL

William F. Mellberg (1952–2017) identified himself on his stationery as "Speaker, Author, Historian." However, he spent his early career, after graduating from the University of Illinois at Urbana–Champaign, in the aircraft industry, working for Fokker, the famed Dutch aerospace manufacturer, as a marketing and public relations representative at Fokker Aircraft Corporation of America, followed by a stint on the marketing staff of Ozark Air Lines and a year trying to get his own aircraft supply parts company, Flight Research, off the ground. At that point Mellberg left the aircraft industry and pursued his avocation as a political humorist and professional speaker. For the rest of his life he addressed trade, professional, and corporate audiences throughout North America, becoming popular for his gentle humor, positive messages, impressions of political figures, and mixing satire with patriotism. He kept active his love for aviation, however, writing stories about flight that appeared in magazines worldwide. He authored two popular books, Famous Airliners *(1995) and* Moon Missions *(1997), the latter a history of lunar exploration with a foreword by Apollo 17 Moon walker, Dr. Harrison H. "Jack" Schmitt.*

LETTERS FROM A GRIEVING WORLD

"HOW NEIL IMPACTED MY LIFE"

Email

August 25, 2012

Carol,

I don't know if anyone is monitoring Neil's emails, but I hope that you receive this. I have been corresponding with Neil the past few months. His most recent email was dated August 12th, just a few days before he went into a coma. I will cherish his note forever! It was from a man that I deeply respect because of the way he lived his life. I have many wonderful "Neil Armstrong" stories that I have shared over the years. I have told them with pride and always stated how Neil impacted my life through his actions. He never used his fame for personal gain or notoriety. He was a private person who valued relationships that were based upon integrity and honesty.

I have photos of Neil and you on my wall. They serve as wonderful reminders of the events where Neil was kind enough to share time with my family and me. One was my daughter's 21st birthday celebration in Oshkosh.

Sharon, Lesley and I express our sincere condolences on Neil's passing. The depth of your sorrow is offset by the heartwarming memories that Neil created, which will be with you for the rest of your life. Each night we look up to the moon, Neil will be smiling back at us.

Sincerely,

Tom
Tom Poberezny
Oshkosh, WI

Neil's email to Tom Poberezny dated Sunday, August 12, 2012, 9:52 p.m., read: "Tom, Many thanks for your kind thoughts. My doctors are not encouraging about my possible attendance in the Citation Jet even for Russ Myer [and his induction into the Citation Jet Pilots Hall of Fame]. However, I will give you a final decision when it is available. All the best, Neil."

Thomas P. Poberezny (b. 1946) is the son of Paul H. Poberezny (1921–2013), founder of the Experimental Aircraft Association (EAA) and its convention and fly-in in Oshkosh, Wisconsin, the largest event of its kind, first staged in 1953.

A RELUCTANT ICON: LETTERS TO NEIL ARMSTRONG

Born into one of America's most avid aviation families, Tom Poberezny became a world champion acrobatic pilot and one of the foremost civil aviation advocates in the United States. In 1977 he took over management of the EAA Fly-In Convention and Sport Aviation Exhibition, better known simply as the Oshkosh Fly-In, by which time was attracting over 12,000 aircraft and more than a million participants and spectators annually. It is not known exactly how many times Neil Armstrong attended the EAA event, but he flew into Oshkosh in his own private plane at least a handful of times. As its director, Tom Poberezny also led the fundraising initiative resulting in the construction of a new EAA World Headquarters and spectacular Air Adventure Museum, the largest private facility of its type in the U.S. As president of the EAA, Tom came to direct an international organization with 750 chapters and over 400,000 members, and which published six different flight magazines. When learning that his condolence letter was going to be published, his wife, Sharon, wrote me to say that "Tom is honored" to have the letter included in this book.

"GAVE FAR MORE TO OUR NATION THAN HE HAD EVER EXPECTED"

August 26, 2012

Dear Mrs. Armstrong,

I want to join the rest of our country in expressing our condolences. Neil was a gentleman, a good man, a good engineer and a good pilot, and he gave far more to our nation than he had ever expected—at a time when we severely needed him and his story.

For you and your family, I am attaching a copy of a photo that was taken at a meeting of the Conquistadores some years ago. Sadly, this week's news brought completion to the story of these three good friends. You probably knew each of these men—Dan Burns, on the left, died this past spring, and my brother, Peter Pfendler, in the middle, died five years ago.

My brother proudly had this photo hanging in his home, and he cherished his friendship with both Neil and Dan.

And, parenthetically, I am a third generation Purdue graduate, grew up in West Lafayette and my father was on faculty. I have been one of thousands of Boilermakers who proudly referred to Neil as an alumnus of our university.

Thank you for sharing Neil with all of us. And thank you for taking a minute to read this letter. We all shall miss him and remember him.

Sincerely,

David F. Pfendler, MD
McMinnville, Oregon

David F. Pfendler was born in 1945 in Lafayette, Indiana, and attended Purdue University. He earned his medical degree from the University of Minnesota and became a specialist in otolaryngology, trained to assess, diagnose, and treat abnormalities of the ears, nose, and throat. He has practiced ENT medicine in McMinnville, Oregon, for some thirty years. David's older brother Peter G. Pfendler (1943–2007) served as a U.S. Air Force pilot on active duty from 1966 to 1970, flying 139 combat missions in Vietnam as a pilot of F-4 Phantom II, receiving 16 combat medals. A 1973 graduate of Harvard Law School, Peter founded Polaris Aircraft Leasing Corporation in San Francisco in 1974. At the time it was the world's largest commercial aircraft leasing company, with airliners leased to twenty-five commercial airlines. In 1989 he sold Polaris to General Electric Credit Corporation and moved to a cattle ranch on Sonoma Mountain, east of Petaluma, California, where he devoted his time to his lifelong love of wildlife conservation and fly fishing. He served on the board of directors of the National Academy of Sciences, The Nature Conservancy in California, and The Peregrine Fund.

The photograph that Dr. David Pfendler mentions in his letter to Carol Armstrong was taken at a retreat of the Conquistadores del Cielo (Spanish for "Conquerors of the Sky"), a highly private club of airline and aerospace industry notables. It is not known when Neil Armstrong first became a member of the Conquistadores, but it was evidently not long after his Apollo 11 mission (the club's history dates back to 1937). Neil often attended the secretive retreats of the Conquistadores, many of which took place on remote ranches in the American West. It is likely that Dr. Pfendler's brother, Peter, was a member of the Conquistadores.

"A GREAT SHOCK AND SURPRISE"

August 26, 2012

Dear Carol—

We only met a couple of times at Golden Eagle affairs but I did know Neil, having played golf with him and dined at the same table several times. With that introduction I want to tell you how sorry my wife Claire and I are at his passing. We, of course, had no idea that he had a problem of any sort when we saw him in Orlando this past April, so his death came as a great shock and surprise. Our sympathy and prayer go out to you and the rest of his family.

I'm sure you have more support than you could ever use, but if this old Naval Aviator can be of any help at all, please call on me.

All the very best,

Bob
VADM Robert Francis Dunn, USN Retired
Alexandria, Virginia

The Golden Eagles comprise an elite association of naval aviation pioneers limited to 200 active members. At one time 5 men from Neil's fighter squadron—VF-51—were members of the Golden Eagles, the most ever from any one squadron.

Raised in Chicago, Robert Francis Dunn (b. 1928) fell in love with aviation when he took his first airplane ride at age ten in 1938. After earning a degree from Northwestern University, he accepted an appointment to Annapolis, graduating from the Naval Academy in 1951. On duty in the Korean War, he served on the escort destroyer USS Nicholas *(DDE-449) then began flight training, earning his aviator's wings in 1953 and flying the AD Skyraider in various attack squadrons. From 1956 to 1960 Dunn was a flight instructor at Pensacola, then served as flag lieutenant to Rear Admiral Joseph "Jumping Joe" Clifton, the legendary World War II navy fighter ace who, in command of Fighter Squadron 12 aboard the USS* Saratoga, *shot down five Japanese airplanes and led the first attacks against Japanese positions in New Guinea. Following shore tours at the Naval Postgraduate School and Bureau of Naval Weapons, Dunn in 1966–1967 flew bombing runs against North Vietnam while serving as executive officer and then commanding officer of Attack Squadron 146 (VA-146). In 1970–1971 Dunn did a short tour as Commander, Carrier Air Wing Seven (CVW-7) before moving to the staff of the Sixth Fleet. Promoted to flag rank in the mid-1970s, Dunn served as Commander, Naval Safety Center; a member of the staff of Commander, Naval Air Force Atlantic Fleet; Commander, Carrier Group Eight;*

Commander, Naval Military Personnel Command; and chief of Naval Reserve. His active service concluded with two tours as a three-star admiral. In 1988 he was honored as the navy's Gray Eagle, an active aviator with the earliest designation. Upon retirement he worked for a time at the United States Naval Institute, for which he had also served as a board member.

"THE WORLD WILL MISS HIM"

August 26, 2012

Dear Carol,

Nathaniel and I convey to you our deepest sympathy on your loss of Neil. The generosity you showed us a few summers back when we dropped by will always be a highlight of our lives. I suppose that single opening of your home epitomizes who Neil was and you are. The world will miss him—we surely will.

If there is anything we can do for you, now or in the future, please do not hesitate to call upon us. You are in our prayers and thoughts.

Sincerely,

Ray Rothrock
Portola Valley, CA

Mr. Rothrock, now retired, was a venture capitalist who in 1997 invested in Space.com. Along with the company's cofounder Lou Dobbs (the television commentator best known for his Fox Business Network program Lou Dobbs Tonight*), Rothrock had the idea to launch Space.com in July 1999, during the 30th anniversary of Apollo 11 and, in building the business (in New York City) thought it would be a good idea to recruit some astronauts to join its board of directors; both Neil Armstrong and Sally Ride accepted the position. According to the email I received from Mr. Rothrock, both he and Dobbs "enjoyed their stories, their wisdom, and their insights into what makes space interesting." Space.com lives on today, though neither Dobbs nor Rothrock is still involved.*

A RELUCTANT ICON: LETTERS TO NEIL ARMSTRONG

"LONG BE REMEMBERED LIKE COLUMBUS"

August 26, 2012

To you all:

It is with deep regret that I learnt this morning of the passing of Neil Armstrong.

    Born in 1930, the same in which I was born, like so many all over the world, I, as a Civil Engineer, admired his accomplishment of being the first to step on a body in outer space, the moon.

    He will long be remembered like Columbus is to us today.

With condolences,

Gerry McGee, P. Eng.
Ottawa, Ontario
Canada

In Canada, P. Eng. is the abbreviation for Professional Engineer (in the U.S. it's PE), a government-certified license to practice engineering in the province or territory where the license was granted, in this case the province of Ontario. It seems likely from the nature of his letter that Mr. McGee had never met Neil Armstrong.

"AN INSPIRATION TO ANYONE THAT MET HIM"

August 26, 2012

Dear Carol,

I am so saddened by such a great loss to all of us and especially to the young of the world and our country. Neil was an inspiration to anyone that met him. His calm confidence, patriotism, wisdom and most especially to me, his deep and sincere humility, epitomized so much of what our times need more of. He is sorely missed. Yet, I know the good Lord has plans for him elsewhere in this great Creation he has made with its deep mysteries that we humans will never surely comprehend in entirety. I believe those mysteries include a way in which we will know our husbands again when our time comes to leave this special earth which is our home. I pray for you in your grief, Carol, and that the legacy of love

Neil has given to you will sustain you in the days and weeks ahead.

This photograph of our son-in-law, David Caldwell's is my very favorite because it reflects so purely God's glory. Come visit in Boca Grande when you can. I will keep in touch through Louise. In the meantime, may God's angels bear you up on their wings.

With love and prayer,

Paula
Paula Lillard (Mrs. John S. Lillard)
Lake Forest, Illinois

Neil and Carol vacationed occasionally in Boca Grande, Florida, on Gasparilla Island in southwest Florida, off Charlotte Harbor near Fort Myers, which is home to many seasonal residents such as John and Paula Lillard.

John and Paula Lillard are active philanthropists, notably through their Red Bird Hollow Foundation based in Lake Forest, Illinois. A U.S. Navy veteran and graduate of the University of Virginia, John cofounded JMB Institutional Realty in 1978. Paula, an educational professional, cofounded Forest Bluff School, a Montessori school in suburban Chicago. In 2015–2016 the couple, via their foundation, gave a seven-figure gift to Lake Forest College, the site of the new Lillard Science Center. Their causes are diverse, involving education, arts and culture, human services and health, and the environment. They also contribute significantly to politically right-of-center policy organizations such as the Cato, Heartland, and American Enterprise Institutes.

"NO LONGER LOOK AT THE MOON IN THE SAME WAY"

Rio, 26th of August 2012

Dear Carol,

I was collapsed yesterday when I head the sad news. Everyone knows the loss to mankind. I guess tributes in this direction are endless.

It is to you that I think at the moment I will remember all my life your eyes and your smile, watching Neil talk in Paris in 2008. The smile that you had sent me to suggest that you'd convinced him to come to Paris . . . This meeting will remain one of the greatest moments of my life.

If you agree, I would like to share the video we made during this keynote, with people who have not had the chance to know him, or who like us, will be pleased to see him talking about this extraordinary adventure.

I think about this letter Neil wrote me to let me share his time to enjoy a 1930 Armagnac the day of his 80 years . . . What attention it was for me!

I already no longer look at the moon in the moon in the same way since we met. But it will be even more emotional now.

Sincerely,

François

PS Please forgive me about this paper. I'm in Rio and I did not want to wait for writing this letter.

François Hisquin
Le Vésinet, France

François Hisquin (b. 1965) is a French businessman who in the late 1990s founded OCTO Technology SA, an IT consulting firm based in Paris. Besides serving as OCTO's chair and CEO, Hisquin has been at the head of five other companies: KPIT Infosystems France SAS, Pivolis SA, Réalisations et Prestations Graphiques, Aubay, and Syntec Numérique. He is also the founder of the USI conferences, an international meeting held annually in Paris since 2008 for the purpose of discovering "Unexpected Sources of Inspiration" through the gathering of "the greatest thinkers, innovative and creative people of our times"[49] for presentations and discourse. Neil Armstrong participated in the inaugural USI meeting in 2008.

"NEIL ARMSTRONG'S BAKED BEANS"

August 26, 2012

Dear Carol,

I was saddened to hear about Neil's passing. We had known each other for many years. Neil and I first met when he was an Engineering Test Pilot at Dryden Flight Test Center. There were times I would be there on a flight program from Ames Research Center and be able to spend time with Neil. Over the years, we saw each other at various NASA sites and, of course, when we both attended the annual SETP Symposium.

He stayed with us on several occasions here at our home in Saratoga. Prior to his lunar flight, he came up to Ames to fly our vertical takeoff and landing experimental aircraft. I was Chief of Flight Operations then and authorized his training with the X-14A in preparation for the lunar landing. It was a rather simple, one-person light aircraft that was modified with a system to provide vertical deflection of the jet and was equipped with jet controls on the wingtips and tails. Thus it became a training device for VTOL flight and served as a vehicle for Neil to get some practice with vertical takeoffs, hovering, and translation modes in preparation for the lunar landing.

Another time, Neil had given a lecture at Stanford University. Afterwards, on the spur of the moment, I invited him to come home with me and have dinner. Louise was caught entirely by surprise. She had only prepared baked beans for dinner—after which the recipe became known in the family as "Neil Armstrong's Baked Beans." The only extra bed we had at that time was a set of bunk beds, which he was able to fit into. Into one of them, that is!

For a time, Neil was Deputy Director of Aeronautics at Headquarters in Washington; worked with Bill Harper. I was in Washington for a NASA/Industry committee meeting with airline VPs, FAA and military service members, and Neil invited me to stay in his apartment. Kindness was part of who he was.

I am 96 now and spend most of my days at home. Over the past decade or so, Neil was very thoughtful in paying a visit when he was in the area. I appreciated his engineering back in the day and his friendship ongoing.

With sincere condolences,

George
George E. Cooper
Saratoga, California

Neil Armstrong considered George E. Cooper (1916–2016) to be one of the greatest test pilots he ever knew—and with very good reason. Cooper's name is not generally known, and he certainly did not fit the popular conception of a test pilot as a glamorous, romantic figure full of devil-may-care bravado and the mythical Right Stuff; rather, he was an expert pilot and dedicated engineer with an analytical mind and keen curiosity for discovering reliable new knowledge about what it took for flying machines and piloting to progress. Born in California and graduating from UC Berkeley—interestingly, in mining engineering—his first work was in the gold mines of the Sierra Madre. The outbreak of World War II, however, took him into aviation, training as a pilot and seeing significant action over Europe: eighty-one missions in the P-47 Thunderbolt with the 412th Fighter Squadron. After the war, he became a test pilot with the National Advisory Committee for Aeronautics (NACA) at Ames Research Center, where by 1957 he had devised what came to be known as the Cooper Pilot Opinion Rating Scale, a seminal contribution to the understanding of airplane flying qualities. Until his retirement in 1973, Cooper served as Ames's chief research pilot. He received almost every award a test pilot could earn, including the Admiral Luis de Florez Flight Safety Award, Richard Hansford Burroughs, Jr., Test Pilot Award, Octave Chanute Award, and Arthur S. Flemming Award. He also received the NASA Career Achievement Award, was named a Fellow of the American Institute of Aeronautics and Astronautics, had the honor of delivering the Wright Brothers Lectureship in Aeronautics, and was inducted into both the National Academy of Engineering and the NASA Ames Hall of Fame. He was also a founding fellow of the Society of Experimental Test Pilots and received the Legion of Honor from the French government for his military service in World War II. He died in 2016 at age one hundred.

"MR. ARMSTRONG'S MEMORY WILL CONTINUE TO BE AN INSPIRATION"

August 27, 2012

Armstrong Family,

I am saddened for your loss. Cdr./Prof. Armstrong was a hero, even if he shied away from such a title, of mine since I watched his moonwalk as a seven-year-old boy. Forty-three years later I still think of him as such.

In 2007 I had a chance encounter with Mr. Armstrong at a Home Depot store in Mason, OH. Out of respect for his cherished privacy, I simply thanked him for his service. He simply nodded thank you and gave a slight, humble smile. I was thrilled. Mr. Armstrong's memory will continue to be an inspiration.

Respectfully,

John Gallardo

P.S. Before I wrote this note tonight, I looked out my apt. window, saw a gibbous moon and winked.

John Gallardo
Cincinnati, Ohio

John Gallardo was born in 1962 in Dearborn, Michigan. Raised in southern West Virginia, he received his undergraduate degree in business administration from Concord College in Athens, West Virginia. In correspondence with me about this book, John wrote that Neil was his "childhood hero."

"SUCH KINDNESS ONE DOES NOT FIND IN THE EVERY-DAY MAN"

August 27, 2012

Dear Carol,

Oh, how to address the passing of our dear and humble friend?
    Our memories are long and personal. Such kindness one does not find in the every-day man.
    Jack and I are blessed to have been touched by your gentle husband.
    Sometime in the future we would love to share our remarkable experiences with Neil.
    We love you, and are keeping you and your family in our on-going prayers.

Fondly,

Sally & Jack
Sally and Jack Chapman
Fort Pierce, Florida

A RELUCTANT ICON: LETTERS TO NEIL ARMSTRONG

Sally Putnam Chapman (b. 1938) is the granddaughter of George Palmer Putnam (1887–1950), the publishing heir (G. P. Putnam's Sons) and, from 1931, the husband of Amelia Earhart. He married the legendary aviatrix in 1931, six years before her mysterious disappearance while flying across the Pacific Ocean. Sally is thus Earhart's step-granddaughter.

Since 1982 Sally has lived with her husband, Jack Chapman, in a remarkable home located west of Fort Pierce known as Immokolee, a word derived from the Seminole word meaning "home place." Built in the Mediterranean revival style and situated inside a hammock of oak trees backed by thirty acres of citrus groves, the home (added to the National Register of Historic Places in 1994) enjoys five bedrooms and six-and-a-half bathrooms; on its grounds are a three-bedroom two-bathroom guesthouse, swimming pool, and exercise room, as well as a six-car garage. The home was built for Sally's grandmother Dorothy Binney Palmer (1888–1982), the daughter of Crayola crayon inventor Edwin Binney, in 1930 following Dorothy's divorce from George Putnam (whom she had married in 1911) and subsequent marriage to Frank Upton, a Congressional Medal of Honor winner from World War I. A friend of Earhart's who sometimes flew with Amelia in her Avro Avian Moth sport plane, Dorothy was herself a tremendously accomplished person—an explorer, mountain climber, naturalist, accomplished athlete and pianist, world traveler, and free-spirited feminist; along the way she became the first woman to climb California's Mount Whitney, which at the time was America's highest mountain. She also became the first woman to swim to the surface of an ocean from a submerged submarine—a feat from World War II when Dorothy opened Immokolee for training of the U.S. Navy's nascent Underwater Demolition Teams, the forerunners of the Navy SEALs.

It is not known when and where Neil Armstrong first met Sally and Jack Chapman, or how regular their friendship became. What is known is that Neil played a vital role in persuading Sally, in 2002–2003, to donate some of her prized papers and artifacts to the archives at Purdue University (which was neither of the Chapmans' alma mater). Foremost in that gift were some 500 letters and assorted items that had belonged to Amelia Earhart. In combination with the 31.1 cubic feet of Earhart documents given to Purdue by George Palmer Putnam back in 1940, Sally's gift (of what Jack Chapman calls some of "the really juicy stuff") strengthened Purdue's status as the single most significant archive of Earhart papers anywhere in the world.

"A GENUINE 'NICE GUY'"

August 27, 2012

Dear Carol:

I want to express my sincere sympathy for your loss. A group of us old WW II veterans had an occasional lunch for Doolittle Raider Tom Griffin, and whenever he was available Neil joined the group, which Tom greatly appreciated. Neil was a genuine "nice guy," and I was honored to include him in my circle of friends. Thank you for sharing him with us—

Sincerely,

Link
Lincoln Pavey
Cincinnati, Ohio

Lincoln Wendell Pavey (1925–2015) was a good friend of Neil and Carol Armstrong who lived, as they did, in the Village of Indian Hill, a southeastern suburb of Cincinnati. A World War II army veteran, Pavey worked as a territory manager for IBM, general manager at Data Entry Services, sales manager at Industry Data Systems, and plant manager at Continental Mineral Processing Company. Later in life, he formed a few corporations of his own, notably a land development company for the construction of residential homes. On the side he hosted The Link Pavey Show *on radio station WMKV and was active in the Cincinnati jazz community. Like Neil, he was passionate about aviation. Not only was he an avid supporter of the Tri-State Warbird Museum in Batavia, Ohio, east of Cincinnati, he was also a member of the Flying Neutrons flying club based at the former Blue Ash Airport located sixteen miles northeast of the city.*

  Neil attended the fiftieth wedding anniversary of Link Pavey and his wife, Francie, in 2012, a couple months before Neil's death. (Carol did not attend the anniversary party because she was visiting her grandchildren in another state.) According to Pavey family members, Link always spoke very fondly of his relationship with Neil and their discussions about their mutual loves, jazz and golf.

A RELUCTANT ICON: LETTERS TO NEIL ARMSTRONG

"A TRUE AMERICAN HERO"

August 27, 2012

Carol,

Words cannot express how sorry we were to learn about Neil's sudden death. It seems like just a short time ago that we saw him with you and a group of Camargo friends having dinner on the porch.

We will cherish forever the times we spent together at Christmas, our dinners together and the amazing birthday party two years ago. I will never forget the honor of rooming with Neil on a trip to Bermuda with the C&C Club—he knew all of the people in the TV news shows personally!

He is a true American hero but you would never realize that from his conversations and warm and friendly interactions. Vivian and I also noted his love for you and love, care and passion for his children and for his friends.

We will think of him every time we look at the moon and will honor him with a "loving wink."

Love,

Vivian & Jim
James E. Schwab
Cincinnati, OH

A graduate of the University of Michigan Law School, James E. Schwab retired in 2016 as president and chief executive officer of Interact for Health, the largest health nonprofit organization in the Greater Cincinnati area. Prior to his five years in that position, he had served as the Cincinnati market president of U.S. Bank, chair of the board of TriHealth and Bethesda, Inc., and a board member of the Health Collaborative and the Center for Closing the Health Gap. In civic affairs he has been active on the boards of the Cincinnati Symphony and Pops Orchestra, Health Policy Institute of Ohio, and Cincinnati Children's Hospital Medical Center Foundation. He also served as trustee of the University of Cincinnati Foundation. With his wife, Vivian, he enjoys the arts and has supported the ballet, symphony, and theater in Cincinnati. A fellow member of the Camargo Club in Indian Hill, he played a number of golf rounds with Neil Armstrong.

"A TRULY GREAT MAN"

August 27, 2012

Dear Carol,

I am saddened by Neil's death and very sorry for your loss.

Neil was a truly great man, not only for what he accomplished, but as much for the modesty and grace with which he handled his fame. And, he was a wonderful person—considerate, friendly, fair—and always a gentleman.

I am privileged and honored to have known him.

Sincerely,

John
John Shepherd
Cincinnati, Ohio

John M. Shepherd was a friend and neighbor of Neil and Carol Armstrong and fellow member of the Camargo Club. Now retired, Shepherd served as chair and chief executive officer of The Shepherd Chemical Company and as director and secretary of The Shepherd Color Company, both based in Cincinnati. He holds a bachelor's degree in ceramic engineering and an MBA from The Ohio State University and was a long-time member of The Ohio State University Foundation Board and trustee of the Ohio Foundation of Independent Colleges.

"HOW FORTUNATE NEIL WAS TO SHARE HIS LATER YEARS WITH YOU"

August 27, 2012

Dear Carol,

I have before me notes written by you and Neil following Bill's death. The last time I saw you both was at Joe Head's 80th, and you came over to sit & keep me company. Our frequent times together made me realize how fortunate Neil was to share his later years with *you*.

I am thankful he did not suffer a long illness or be an invalid. Your family message was perfect—I still wink at the moon!

Fondly,

Jean Wommack
Cincinnati, OH

Another person with whom Neil and Carol Armstrong became friends through their mutual memberships at Camargo Club was Jean Emery Wommack (1926–2014). She was an accomplished golfer, having learned to play as a teenager from legendary touring pro Byron Nelson at the Inverness Club in Toledo. Neil and Carol played golf frequently with Jean and her husband, Bill, at Camargo. A graduate of Wellesley College, Jean taught school in New York City, Toledo, and Cincinnati, and for many years she provided volunteer music therapy in the Cincinnati public schools.

"WE HAVE FRIENDS IN HIGH PLACES"

August 27, 2012

Dear Armstrong Family,

I am saddened at learning of the passing of your beloved Neil. I lost my own dear husband 8 years ago. He also worked for NASA. I was living in Satellite Beach in 1969 and watched Apollo 11 lift-off—how amazing that was! The following July I gave birth to my first son in Melbourne, Florida. He was nameless for a few days until my doctor suggested we name him Neil since they were both "launched" from Brevard County (a year apart). My Neil has always been proud of being named for your Neil. Although we have never met him, we did get a few feet away from him at a function here in Huntsville.

One of the things that comforted me after my husband died was a NASA poster saying "We have friends in high places." Knowing our loved ones are with God in High Places helped.

May you find comfort in the love that surrounds you.

Yours truly,

Suzy Szymczak
Huntsville, AL

As Susan Marie "Suzy" Szymczak (1946–2014) wrote in her condolence letter, her husband worked for NASA at Kennedy Space Center in the 1960s. Indeed, a number of Szymczak family members over the years worked in the aerospace industry and NASA. In her later years Suzy was an active member of the Red Hat Society, an international society of women age fifty and older dedicated to reshaping the way women are viewed in today's culture. Founded in 1998, the society embraces the five Fs: fun, friendship, freedom, fulfillment, and fitness. The Huntsville association of Red Hat Rovers has been one of the most active of the some 20,000 Red Hat Society local chapters nationwide.

"THE HIGHEST PLACE IN THE CLUB'S AFFECTIONS"

August 27, 2012

Dear Mrs Armstrong

We were so very sad to hear the news of your husband's death, and on behalf of The Captain, Members and Staff of The Royal and Ancient Golf Club of St. Andrews, may I please send you and all the Family our very deepest sympathy.

It was such an honour for the R & A to count Neil as one of its Members, following his election to the Club in 1986. When he retired from membership in 2003, he continued to enjoy the highest place in the Club's affections and will be enormously missed by all his friends in St. Andrews.

Our thoughts are with you all at this time of great loss, and we will certainly heed your wonderful wish for a thought and a wink whenever we glance at the Moon.

Yours sincerely

Peter Dawson
Secretary
The Royal and Ancient Golf Club of St. Andrews
Fife
St. Andrews, Scotland

Peter Dawson served for sixteen years (1999–2015) as the chief executive of the Royal and Ancient Golf Club of St. Andrews, the body that stages the [British]

Open Championship and governs the sport of golf worldwide in conjunction with the United States Golf Association. Prior to his time with the R&A, Dawson, a British citizen, worked for Grove Worldwide, an American company that manufactures cranes and earth-moving equipment, running the company's operations in Europe, Africa, and the Middle East. Born in Aberdeen in 1948 and raised in Edinburg, the Scotsman perhaps surprisingly was not part of a golfing family. Nonetheless, he became a scratch golfer and captained the Cambridge University golf team in 1969 while studying for a degree in engineering. At least one of the rounds that Neil Armstrong played on the Old Course at St. Andrews was in the company of Peter Dawson.

"PRIVILEGE FOR US TO HAVE KNOWN HIM"

August 27, 2012

Dear Carol,

A television announcement during the past weekend brought us the sad news of Neil's death. Charley joins me in sending heartfelt sympathy to you and to all of Neil's family.

Of course, you know better than any of us that Neil was a man of the highest caliber. He was very bright, courageous, accomplished in his field, a real hero and a truly modest man. It was a privilege for us to have known him and you and I am grateful to have had fun with you both at Eaton events we attended together.

Over the weekend Charley and I recalled being with you and Neil on the terrace at Bald Peak last summer when you were visiting friends in New Hampshire. It was kind of your friends to arrange our meeting—I know it meant a lot to Charley to spend some time with Neil, a man he very much admired.

There probably is nothing I might write to comfort you at this time but I hope that the memories of the many good years that you and Neil had together will soon give you the strength to cope with your loss.

With warmest regards,

Nina
Nina Hugel
Melvin Village, New Hampshire

Cornelia F. "Nina" Hugel (1929–2019) and her husband, Charles E. Hugel (1928–2016), were friends from Ohio who in retirement lived in New Hampshire in the summer and Gulf Stream, Florida, in the winter. Carol and Neil visited the Hugels in both places. In New Hampshire the Hugels resided for many years at the Bald Peak Colony Club, which featured a classic Donald Ross golf course (designed in 1922) that was one of the finest links in all of New England, and which became one of Neil's favorites to play. (For several years Charley Hugel was chair of the PGA Senior Tour, and Nina, too, loved to play the sport.)

Many friends have suggested that Nina—a graduate of Smith College—was "the woman behind the man" in their sixty-four-year marriage. Without question, Charley had an outstanding career, leaving AT&T in 1982, after thirty years, as the executive vice president responsible for AT&T International, Bell Laboratories, and Western Electric Company. After retiring from AT&T, he immediately took the position of CEO of Combustion Engineering, later becoming its chair of the board. During his ten years there, he was responsible for negotiating the first joint venture between a U.S. company and the Soviet Union. He would later serve as executive chair of the board for RJR Nabisco, where he presided over the leveraged buyout of the corporation by Kohlberg Kravis Roberts, an infamous episode in the history of American business that was documented in the book and associated motion picture Barbarians at the Gate. *Charley served as the chair of the U.S./Soviet Trade Commission and the U.S./China Trade Council.*

The Hugels moved from Bald Peak Colony Club to the Sugar Hill Retirement Community in 2015, the year before Charley's death.

"HIS LEGACY IS BOTH EXTRAORDINARY AND ASSURED"

August 27, 2012

Dear Mrs. Armstrong,

On behalf of the entire University of Southern California community, I want to express our deepest condolences on your husband's passing. We are certain he was a source of tremendous inspiration and joy for you, and that you will always cherish your memories of the special times you shared. At USC, we felt tremendously privileged to call Neil one of our own. It was so wonderful to welcome him back to his alma mater as the speaker at our 2005 commencement exercises. His presence made that event even more momentous for the graduating students and their families, and we all have fond memories from that very special day.

As one of our nation's most courageous explorers, Neil certainly earned an esteemed place in history. He achieved so much over the course of his illustrious career. In addition to his professional achievements, he will always be remembered for his great humility and intelligence, his unwavering pursuit of excellence, his exceptional fortitude, and his dedicated service to our country. His legacy is both extraordinary and assured.

You and your entire family remain very much in our thoughts, Mrs. Armstrong. Please accept our heartfelt sympathies.

Yours truly,

C. L. Nikias
C. L. Max Nikias
President
University of Southern California
Los Angeles, CA

Chrysostomos Loizos "Max" Nikias was president of the University of Southern California from August 2010 to August 2018. A veritable Renaissance man, prior to his presidency he held faculty appointments at USC in both electrical engineering and the classics, and he regularly taught a popular undergraduate course on the culture of Athenian democracy (he was a Cypriot American born on the island of Cyprus in 1952 and educated at the National Technical University of Athens and the University of Buffalo).

A bronze statue of Neil Armstrong as an astronaut, with his helmet under his right arm, stands in the quad of the Vieterbi School of Engineering at the University of Southern California. Neil received a master's degree in aerospace engineering from USC in 1970.

"HIS CALM LEADERSHIP & SENSE OF PURPOSE"

August 27, 2012

Carol,

It's hard for me to adequately express my upset when I heard of Neil's death. I didn't know him as a national hero. To me, he was the engineer, the professor, the business associate, and, most importantly, the friend.

No one will note that Neil personally saved an important defense resource AIL and, with it, thousands of jobs. But I vividly recall his calm leadership & sense of purpose.

Please accept my heartfelt condolences.

Steve
Stephen R. Hardis
Moreland Hills, OH

One of the first corporate boards onto which Neil Armstrong agreed to serve, starting in 1979 (and lasting into the early 2000s), was that of the Eaton Corporation, based in Cleveland, Ohio. That same year Stephen R. Hardis (b. 1935) began his career at Eaton as executive vice president for finance and administration. Moving up the ranks, in 1986 Hardis was elected Eaton's CFO and from 1996 to 2000 its CEO and chair. Founded in 1911 in Bloomfield, New Jersey, Eaton moved to Cleveland in 1941 to be closer to what was then its core business, the automotive industry. Today Eaton Corporation, Inc., is a multinational power management company with its corporate headquarters in Dublin, Ireland, and its operational headquarters in Beachwood, Ohio, a Cleveland suburb. In 2018 it did $21.6 billion in sales in three major sectors: hydraulics, aerospace, and vehicles.

Beyond his work for Eaton, Hardis, a Phi Beta Kappa graduate of Cornell University with a master's degree in public and international affairs from Princeton University, also served as director of Axcelis Technologies, Inc., American Greetings Corporation, Lexmark Corporation, Marsh & McLennan Companies, Inc., Nordson Corporation, Progressive Corporation, and STERIS Corporation. He was also a long-time board member of the Cleveland Clinic Foundation.

"WORDS ARE SO INADEQUATE"

August 28, 2102

Dear Carol,

Words are so inadequate. We are all so devastated by the loss of our squadron buddy and world hero. There will never be another Neil. History will list him right above Lindbergh & Columbus.

No one should have to go through this twice. You became my very cherished friend at the 2001 reunion when you offered consoling words to me after the loss of Betty. I will never forget your kind words.

God be with you, Carol. You are in my prayers.

Love,

Bob Kaps
VF-51
Wellsville, KS

Robert J. "Bottle" Kaps (1928–2012) was a naval aviator who flew in the Korean War with Neil Armstrong as part of Fighter Squadron 51 (VF-51), flying 103 combat missions. He died on December 17, 2012, some three and a half months after Neil's death. Raised on a farm in north central Kansas, Kaps, upon returning from Korea, took a job with the Federal Aviation Administration at the Russell Airport located not far from his home. In 1972 he moved with his family to Kansas City, where he took a post at the FAA Central Region administration office. Kaps finished a twenty-nine-year career in the civil service in 1985. A life member of the American Legion and the Veterans of Foreign Affairs, he was an ardent supporter of veterans' rights and causes. He regularly attended the reunions of his VF-51 squad members, thereby staying in close touch with Neil.

"HIS WONDERFUL UNPRETENTIOUSNESS"

August 28, 2012

Dear Carol—

We were so deeply saddened to hear the news of your husband's passing on Saturday. The two of you flew to Dallas about 4 years ago to have dinner with our family. It was a special and historic day for the five of us, especially our young children to meet Neil and have the opportunity to hear some of his stories about his life. Your husband was a true hero and such an admired man not only for his accomplishment, but almost even more for his modesty, his humility, his patriotism, his aspirations and his wonderful unpretentiousness. Our children will always remember this about him and we hope will honor this legacy by living life the same way. The world will miss Neil Armstrong, but his heroic legacy will live on for time eternal. Our love and prayers are with you and your family at this very difficult time.

Yours Truly—

Kathy & Harlan Crow
Jack, Rob, and Sarah
Dallas, Texas

Harlan Rogers Crow (b. 1949) is a real estate developer from Dallas. From 1974 to 1978 he worked as a leasing agent for Trammell Crow Houston Industrial—a business started by his father, Trammell Crow (1914–2009), the real estate developer responsible for the Dallas Market Center, Peachtree Center in Atlanta, and Embarcadero Center in San Francisco—subsequently managing the Dallas office building development operations of his father's company until 1986. Harlan Crow then served as president of the Wyndham Hotel Company before assuming responsibility for Crow Holdings in 1988 and thereafter serving as its chair and CEO. Politically he has been highly active, donating to Republican campaigns and conservative causes and serving for many years on the board of the American Enterprise Institute. He is a close friend of Supreme Court Justice Clarence Thomas and a member of the controversial all-male Bohemian Club. (Neil Armstrong was never a member of the Bohemian Club, but he did on a couple of occasions accept an invitation to spend a few days at the club's annual summer encampment at the Bohemian Grove in Northern California.) Crow's home in Dallas holds an extraordinary collection of historical materials involving the likes of Ponce de Leon, Christopher Columbus, Amerigo Vespucci, George Washington, Napoleon Bonaparte, the Duke of Wellington, Frederick Douglas, Abraham Lincoln, and Robert E. Lee, as well as all the signers of the Declaration of Independence and U.S. Constitution. In his collection is also art by Renoir and Monet as well as paintings by Winston Churchill and Dwight Eisenhower. In his backyard garden stands a series of sculptures featuring historic figures that Crow considers to be failed or fallen leaders, including Vladimir Lenin, Josef Stalin, Fidel Castro, Che Guevara, Hosni Mubarak, Josip Broz Tito, Nicolae Ceausescu, and Bela Kun. Crow acquired these former public monuments in the early 1990s after the collapse of the Soviet Union and Eastern Bloc countries.

A RELUCTANT ICON: LETTERS TO NEIL ARMSTRONG

"THE KIND OF TRUE PATRIOT WE ALL ASPIRE TO BE"

August 28, 2012

Dear Mrs. Armstrong,

It is with great sadness that I learned of the passing of your husband. On behalf of the Naval Air Forces, please accept my sincere condolences.

Words are inadequate to provide consolation, but please know you and your family are in our thoughts and prayers. Neil Armstrong was a skilled pilot, a humble yet engaging leader and the kind of true patriot we all aspire to be. He served his country with the utmost honor, and we are all proud of his many achievements and contributions to our Nation. He will always be an icon of Naval Aviation.

As your family gathers to celebrate your husband's life, remember your Navy family stands with you in prayer. Please let us know if there is any way we can be of assistance to you.

Sincerely & respectfully,

A.G. Myers
Vice Admiral, U.S. Navy
Commander, Naval Air Forces

A 1978 graduate of the U.S. Air Force Academy who grew up in northern Virginia, Vice Admiral Allen G. Myers (b. 1956) served as Commander, Naval Air Forces, from 2010 to 2012. In that post, he was in command of all U.S. Navy naval aviation units. Previous to that, among other tours of duty on the sea and in the Pentagon, Myers served as commander of Carrier Strike Group Eight, commanding the Expeditionary Strike Force 5th Fleet, Combined Task Force 50 and 152, and the Eisenhower Carrier Strike Group during an extended deployment in 2006–2007 in support of Operations Iraqi Freedom and Enduring Freedom. Earlier squadron and sea tours for him included VF-143, VF-14, VF-101, and VF-103. Prior to his retirement from the navy in 2014, Admiral Myers was deputy chief of Naval Operations (Integration of Capabilities and Resources) in Washington. During his career he has accumulated more than 3,600 flight hours and over 900 carrier landings. His distinctions and military decorations include many of the country's most distinguished awards and are too numerous to list.

LETTERS FROM A GRIEVING WORLD

"A SENSE OF CALMNESS & STRENGTH WOULD FILL ME"

August 28, 2012

Dear Carol,

I could not possibly add to any of the tributes you are reading . . . but I do want you to know that I am there with you in your circle of love & support as you were for me.

 So many happy times we shared—knowing you as a loving, caring couple.

 You always knew when I needed a friend at my side after Jim died. The two of you would come over & stand at my side—Neil would give my hand a squeeze & a sense of calmness & strength would fill me. I know you have that.

Grateful for all your friendship
With Love,

Chris
Chris Geier
Cincinnati, Ohio

Christine Paske Geier (b. 1947) died on October 15, 2012, less than two months after writing this condolence letter for Neil's passing. She and her previously deceased husband, James A. D. Geier (1925–2001), lived in the Village of Indian Hill as did the Armstrongs. James Geier was chair and CEO of Milacron, Inc., during some of the years that Neil had served on the company's board of directors.

"WE'RE ALL BETTER FOR HAVING KNOWN HIM"

August 29, 2012

Dear Carol:

It is with deep regret that I've received the news of Neil's untimely passing. I have no doubt this is a terribly difficult time for you, but please know that our thoughts are with you during this challenging period.

 For those countless lives Neil touched, we're all better for having known him. I feel privileged for the time I spent with Neil over the last ten years, at Barron's, the Conquistadors, and on so many other occasions. Every conversation was always inspiring and enriching. Perhaps most of all, I am grateful for his friendship.

Neil's tireless advocacy for a forward-thinking space policy was courageous and visionary. For the company I lead, we are grateful for his leadership in future space exploration objectives. The very essence of the human spirit's desire for discovery is the vision that Neil expressed. At EADS, we will redouble our commitment to capture a fraction of his passion to explore.

With deepest condolences,

Tom Enders
Chief Executive Officer, European Aeronautics and Defence and Space (EADS)
Blagnac, France

Dr. Thomas Enders (b. 1958) is a German corporate executive and politician who is currently serving as president of the German Council on Foreign Relations (DGAP), a private, independent, and nonpartisan body that advises the German government on its national foreign policy. From 2012 to early 2019 he was the chief executive of Airbus and its earlier incarnation as the European Aeronautic Defence and Space (EADS) company. Enders holds a doctoral degree in political science from UCLA. His work experience includes time with the German Parliament, International Institute for Strategic Studies in London, German Federal Ministry of Defence, DaimlerChrysler Aerospace, German Aerospace Industries Association (BDLI), and the presidium of the Federation of German Industries (BDI). In his capacity as chief executive of EADS/Airbus, Enders was appointed to British prime minister David Cameron's United Kingdom Business Advisory Group.

"NEIL INSPIRED THE WORLD AND HE INSPIRED ME"

August 29, 2012

Carol,

It was a great privilege to be able to spend time with Neil and you at Purdue. Neil inspired the world and he inspired me.

Sully
Chesley Sullenberger
[Address withheld]

Chesley Burnett "Sully" Sullenberger (b. 1951) became internationally renowned on January 15, 2009, when he and his crew safely guided US Airways Flight 1549 to an emergency water landing in New York City's Hudson River. (His Airbus A320's two engines had lost thrust following a bird strike.) Born and raised in Denison, Texas, Sullenberger graduated from the U.S. Air Force Academy in 1973, receiving the Outstanding Cadet in Airmanship Award. In the air force he served as a fighter pilot, flight leader, and training officer, attaining the rank of captain. In 1980 he became an airline pilot with Pacific Southwest Airlines, later acquired by US Airways. He retired from commercial flying in 2010. His book Highest Duty: My Search for What Really Matters *(2009) became a New York Times Best Seller, and in 2012 he published a second book,* Making a Difference: Stories of Vision and Courage from America's Leaders. *Director Clint Eastwood made a major motion picture about Sullenberger's life,* Sully, *starring Tom Hanks in the lead role. Today Sullenberger lectures internationally at educational institutions, corporations, and nonprofit organizations about the importance of aviation safety and patient safety in medicine, high-performance systems improvement, leadership, crisis management, lifelong preparation, and living a life of integrity.*

"OUR HEARTS ARE BROKEN"

August 29, 2012

Dear Carol,

Della and I are deeply saddened to learn of Neil's passing. Our hearts are broken and we will miss him a great deal.

As you deal with your healing process, if there is anything that Della can do, please let us know. We are truly sorry.

Ernie & Della
Ernie and Della Green
Kettering, OH

A RELUCTANT ICON: LETTERS TO NEIL ARMSTRONG

Ernest "Ernie" Green (b. 1938) is best known for being a running back with the Cleveland Browns from 1962 to 1968. Born and raised in Columbus, Georgia, Green was a college All-American at the University of Louisville. Not until many years later did Neil Armstrong meet and become good friends with Ernie, when he and Green served together on the board of directors for Cleveland's Eaton Corporation. Following his football career, Green worked at Case Western Reserve University as its assistant vice president for student affairs and then as executive director and vice president of the team sports division for International Management Group, a leading sports, events, and talent management company. In 1981 he left IMG to establish EG Industries (EGI), based in Dayton, Ohio, which manufactured components for the automotive industry and later expanded into making parts for medical, consumer, energy, and industrial devices. Besides sitting on Eaton's board, he also served as a director on the board of Dayton Power & Light. Speaking to me from his home in Kettering, a suburb of Dayton, Green remarked with great fondness for Neil, "We were very tight."

"ONE OF A KIND"

August 29, 2012

Dear Carol,

Neil was one of a kind—a true American pioneer and hero. I have never known anyone who has done something so heroic and history-making yet was so self-effacing about it. His entire career, even before Apollo, was admirable. I was privileged to have served on the Milacron board with him.

A fond memory is that board trip to Italy years ago. We were all at a restaurant on a hill above Lake Como. Seating was family style. My husband Wally Barnes was seated at the end of a long table—Neil was sitting further down the line. Wally was surprised when the waiters and chefs kept coming to get photos taken with him. Finally it dawned on us that they thought he was Neil Armstrong! As I recall, Neil couldn't have been more amused.

We mourn Neil's loss. We were honored to know him as a colleague and friend and send deepest sympathy and love to you and your family.

Barbara

Winking at the moon is a great idea!
Barbara Hackman Franklin
Washington, D.C.

Although Barbara Hackman Franklin held a number of political posts with Republican presidential administrations starting with Richard Nixon, Neil Armstrong knew her best in association with her many activities in the business and corporate world. In the private sector Franklin served on the boards of fourteen public companies, including Aetna, Inc., Dow Chemical, Westinghouse, and Nordstrom, as well as four private companies. She served as chair of the National Association of Corporate Directors and in December 2014 was inducted in the NACD Directorship Hall of Fame. She also chaired the Economic Club of New York, presided over the Management Executives' Society, and was a board member of the U.S.-China Business Council, National Committee on U.S.-China Relations, Atlantic Council, Nixon Foundation, and National Symphony Orchestra. She was a founding member of Executive Women in Government (in 1973) and the Women's Forum of Washington, D.C. (1981). In 1971 she led the first White House effort to recruit women for high-level government jobs as a staff assistant to President Nixon, an effort that resulted in nearly quadrupling the number of women in those positions. Following this, President Nixon appointed her an original commissioner of the U.S. Consumer Product Safety Commission, where she focused on safer products for children. Her other political posts have included four terms on the Advisory Committee for Trade Policy and Negotiations and U.S. Secretary of Commerce for President George H.W. Bush; during the latter she achieved a major goal—increasing American exports, with emphasis on market-opening initiatives in China, Russia, Japan, and Mexico. Altogether, Franklin has served five U.S. Presidents. In 2006 she received the Woodrow Wilson Award for Public Service. In 2017 Time magazine named her one of the 50 Women Who Made American Political History.

"HIS QUEST TO UNDERSTAND"

August 30, 2012

Carol,

We were so saddened to hear of Neil's death.

In 1997, it was wonderful to get to know you and once again work with Neil while visiting the World Air Games in Turkey. So many good memories of flying around the country (remember when Neil noted the defective starter on the old German bomber) but I especially recall the evening Neil gave us an impromptu sky tour as we lay back on the tables at that outdoor restaurant someplace in the middle of Turkey.

I'm sure that you are flooded with notes and letters. Just wanted to let you know we share your loss in a small way.

During the filming of "First Flights" I appreciated how Neil went out of his way to introduce me to many of his former colleagues. I came to deeply appreciate his quest to understand, discover and his desire to enable others to share his fascination with learning.

We appreciated the notes and emails from Neil over the intervening years, a real testimony to his character and principles. He is missed.

Mark and Donna Tuttle
Harleysville, PA

Mark Tuttle (b. 1950) was the executive producer (PMT, Limited) for the television documentary series First Flights, *hosted on-air by Neil Armstrong. The program ran for three seasons (thirty-nine episodes), 1991–1994, on the Arts & Entertainment Network. In the series, Armstrong interviewed fellow pilots and aerospace engineers and discussed the history of flight, from hot-air balloons to spaceflight.*

"A REMARKABLE MAN AND A UNIQUE PERSONALITY"

August 30, 2012

Dear Carol,

As you know better than anyone, Neil was a remarkable man and a unique personality. His achievement was as singular as his way of dealing with it in retrospect.

It took only a few minutes at Tenerife to realize how fortunate he was to find a partner who could support him through the aftermath of fame and the end of life with so much love and grace.

You are a wonderful woman, and I wish you happiness and a safe passage through the travails of grief. Please let me know if my own experience of loss can be any help to you. And of course, it would be my pleasure to see you when you come out to visit Andy's family.

Neil was one of very few men who were privileged to live out such a great voyage: *perardua ad astra*.

Love and Sympathy,

Mort
Dr. Morton Grosser
Menlo Park, CA

Born in 1931 in Philadelphia, Morton Grosser met Neil Armstrong on a number of occasions, the last being in 2011 when they both served as delegates to the Starmus World Astrophysics Congress, held on Tenerife, Canary Islands. A recipient of BS and MS degrees in engineering from MIT and a PhD in the history of science from Stanford University, he has taught at both universities and lectured at Dartmouth College, Santa Clara University, UCLA Medical Center, and University of Texas at Austin, as well as several schools in other parts of the world. He is pilot (from age sixteen), an inventor with multiple patents, and the author of nine books, including The Discovery of Neptune *(1962),* Diesel: The Man and the Engine *(1978),* Gossamer Odyssey: The Triumph of Human-Powered Flight *(1981),* On Gossamer Wings *(1982), and* 100 Inventions That Shaped World History *(1993). Dr. Grosser, in fact, codesigned and built components for the three human-powered airplanes, including the Gossamer Albatross. At age eighty-eight, Dr. Grosser is still active, teaching at the Stanford Biodesign Institute and on the faculty of the National Venture Capital Institute at Emory Institute in Atlanta.*

"NEVER FORGET THE SPARKLE IN NEIL'S EYES"

August 31, 2012

Carol,

I have wonderful memories of Neil, including his calm demeanor and effective leadership while chairing Milacron's audit committee.

Most of all, I recall the dinner we all shared at the restaurant in Italy on the occasion of Neil's retirement from the Milacron board. You may remember the waiters coming to our table pestering Wally Barnes for his autograph, of course mistaking Wally for Neil. I'll never forget the sparkle in Neil's eyes and his subtle grin as we all enjoyed Wally's newfound fame.

Truly, Neil will be missed.

Sincerely,

Bob Lienesch
Bonita Springs, FL

Robert P. Lienesch (b. 1946) was chief financial officer of Milacron, Inc., in Cincinnati from 1999 to 2005, following his service as vice president of finance and treasurer. Lienesch joined Milacron in 1979 as director of corporate accounting. In 1981 he served as controller of the robotic and electronic products groups. Prior to joining Milacron, he worked as an audit manager for Ernst & Young.

"I DID FEEL HIS PRESENCE"

August 31, 2012

Dear Carol:

I just returned to my California home after a several hour non-stop flight during which I had the opportunity to reflect on the events of today. Throughout the cross-country flight I was being followed by a beautiful full moon which I enjoyed watching out of my passenger window. That same moon is now shining in my office window as it is a very clear night.

You and the rest of the family put together a wonderful tribute to Neil today. The location and setting were perfect, the speakers were memorable, and even the weather cooperated. The sad part was that Neil was not there in person but I did feel his presence. I am confident that he would have fully approved of it all.

Thank you very much for the photos and the airplane model. The model is especially meaningful for me since it is the airplane I currently fly for United. I am sure that it was given to Neil as part of his service on the Board of Directors of United Airlines. His influence is still felt there today.

If there is any other aviation related stuff that you want to get rid of after the Purdue people are there, please let me know. Also, please let me know if I can help with the transportation to Washington, DC.

You and Neil have always been so kind to me and I really appreciate your friendship. I am so glad that I was able to attend and be with you all today. Please let me know if there is anything I can do for you at anytime.

Your Friend,

Lynn M. Krogh
Seal Beach, California

Neil knew Lynn Krogh (b. 1952) in Krogh's role as a senior captain for United Airlines and as cofounder of International Jet Aviation Services. A magna cum laude graduate of Arizona State University, Krogh came to possess extensive flight experience in a variety of commercial and general aviation aircraft, including gliders and seaplanes. He owned and flew his own Cessna 172 for recreation and to introduce others to aviation. Krogh served on the steering committee for Wings Over the Rockies, a Colorado-based nonprofit dedicated to educating and inspiring all people about aviation and space endeavors of the past, present, and future. In its programs Wings Over the Rockies utilized both the Air & Space Museum in Denver and the Exploration of Flight center at Centennial Airport in Englewood, Colorado.

"FELT LIKE YET ANOTHER PART OF MY CHILDHOOD HAD DISAPPEARED"

August 31, 2012

> "A man's departure from his familiar world may be inevitable, but his spirit lives on, in the deeds and actions of those who remain, in the memories of those left behind, his friends and family, whose lives may reflect the lessons they have learned from him, and that shall become his truest legacy."
>
> —Alan Furst, 2000

Dear Carol,

As I look up at tonight's Blue Moon, I am moved to recall how the events of 7/20/69—which I shall never forget—changed how many of us thereafter viewed the world. Your husband will forever be recalled in the same breath as Marco Polo, Columbus and Magellan. I learned of his passing while helping my daughter move into her dorm room and, when I heard the sad news, tears filled my eyes. I felt like yet another part of my childhood had disappeared.

Gary Slater
Federal Way, WA

Gary R. W. Slater (b. 1953) is a retired attorney who specialized in estate planning, wills, and personal injury.

"COME ALONG ONCE IN A LIFETIME"

August 31, 2012

The Neil Armstrong Family,

Just a note to express my sadness at the loss of your father & husband. Neil Armstrong was a "True American Hero" to me & countless others. Men like Neil Armstrong come along once in a lifetime & I'm letting you know my heartfelt & sincere appreciation for the service Neil Armstrong gave to The United States of America, both in Korea & at NASA. If more intelligent, classy & humble men like Neil Armstrong were running our country & led us by example America would have always remained the Proud, Brave, & classy country it once was!

God Bless Neil Armstrong & your family—I will miss him, & never be able to repay him!

Respectfully sent,

Paul Goresh
West Caldwell, New Jersey

Neil did not know Paul Goresh, nor did Carol Armstrong know who Goresh was when she received his condolence letter in 2012. But Goresh (1959–2018) was himself a reluctantly famous man, the amateur photographer who, on December 8, 1980, snapped the only photo of former Beatle John Lennon with his killer, Mark David Chapman, outside the Dakota apartment building at Central Park West at 72nd Street in New York City, just hours before Lennon's murder.

The story of Goresh's photograph of Chapman with Lennon was covered by several newspapers and other media following Goresh's death after a long illness in January 2018; for example, see Graham Rayman and Ginger Adam Otis's January 16, 2018, article in the New York Daily News, *"Paul Goresh, Who Got Only Picture of John Lennon with His Killer Hours before the Beatle Was Shot, Has Died."*[50]

"THE SERVICE WAS SO APPROPRIATE"

September 1, 2012

Dear Carol,

All the words have been said and there is no way to express our sadness. As we were standing on the 9th fairway after Neil's service, I was looking around at all of your and Neil's friends and thinking how much Neil would have loved being there and how much it would have meant to all of us to have had him with us. The service was so appropriate and the best was having Piper read the 23rd Psalm.

Having the Neil Armstrong New Frontiers Initiative fund at Children's is a wonderful reminder to us all of Neil's extraordinary contributions and of his warmth.

Carol, we are thinking of Neil and of you. Shannon joins me in sending our love.

Lee Ault Carter
Cincinnati, Ohio

Lee Ault Carter (b. 1938) and his wife, Shannon Kelly Carter (b. 1949), were good friends of the Armstrongs, with both couples deeply involved in Cincinnati-area philanthropy, notably with the Cincinnati Children's Hospital Medical Center. A native Cincinnatian who graduated from Princeton University, Lee's career was in marketing, first with the Drackett Company and then with Local Marketing Corporation, a national marketing consulting firm he founded in 1971 whose clients would come to include Procter and Gamble, Quaker Oats, Lever Brothers, Coca-Cola, and S. C. Johnson. Shannon owned the popular Hyde Park Shop for high-end women's shoes, clothing, and accessories. Both together and individually, the Carters have been among Cincinnati's most active civic benefactors. Lee's involvements include being chair of the board of Cincinnati

Children's Hospital, chair of the Fine Arts Institute, chair of the Urban Design Review Board, plus being on the board of the Cincinnati Art Museum and Cincinnati Symphony Orchestra. He also founded and chaired the Cincinnati Arts & Technology Center, a program that engaged over 2,500 at-risk Cincinnati public school juniors and seniors. Shannon was a member, most notably, of the Leadership Cincinnati Class 19 group that launched Crayons to Computers, a free store for teachers (since 1997 replicated in forty-two other cities around the country) that has distributed over $100 million worth of school supplies to teachers serving 100,000 students in Greater Cincinnati. Along with being named Enquirer Woman of the Year by the Cincinnati Enquirer, she was inducted into the Ohio Women's Hall of Fame.

"NEIL WAS EVERYONE'S HERO"

September 1, 2012

Dear Carol,

Since learning of Neil's death on Saturday afternoon I have been broken hearted for you. Other than my own dear husband, I never saw a husband and wife more adoring of one another than you and Neil. Each time we were together at Golden Eagles to see Neil's "glow" when he walked with you. When Ed was inducted into the Golden Eagles Ed was so honored as I am sure Neil was too. But I loved Ed's comment to Neil, "Everyone has heard of Neil Armstrong, few have ever heard of Ed Clexton!"

Carol, I know you believe this, but I'll say it anyway, how fortunate you are to have had the time and marriage you had with Neil, even tho it was never long enough. I feel for you—like you were robbed with not enough years together. Our Lord does have a plan and I just hope and pray you can find comfort through the Lord. That is so tough, but it is the only way of coming through such a loss, with Our Lord guiding you. Of course we didn't know about Neil's heart surgery and post op and we are so sorry and sad.

Ed and I extend our deepest sympathy over your loss and we hope you will always feel welcome to attend any year, whenever future Golden Eagles reunions. Neil was everyone's HERO and your loss is our loss.

With our deepest sympathy and God Bless you,

Catherine & Ed Clexton
Virginia Beach, VA

Edward William Clexton Jr. (b. 1937) is a retired naval aviator, test pilot, and vice admiral whose career in the U.S. Navy ran from 1955 until his retirement from active duty in 1993. He was the first executive officer and third commanding officer of the USS Dwight D. Eisenhower *(CVN 69), an aircraft carrier commissioned in 1977. He made three combat tours during the Vietnam War, flying the F-4 Phantom jet, a plane that earlier he had flight-tested in joint programs involving the Royal Navy/RAF and USN/USAF. During his career as an aviator, Clexton was in the air for over 4,000 hours, for which he received twelve Air Medals and two Navy Commendation Medals (with combat V). His father, Edward William Clexton (1900–1966), was also a vice admiral and a Naval Academy graduate from Annapolis (Class of 1924; his son Class of 1960). Ed Sr. was awarded the Navy Distinguished Service Medal and Legion of Merit; from 1956 to 1960, he was in charge of the U.S. Navy's overall production, procurement, and contracting policy operation in the Pentagon. Neil Armstrong respected the Clextons as one of America's greatest navy families.*

"SO KIND AND GENEROUS WITH HIS TIME"

September 1, 2012

Dear Carol,

We were so sorry to hear about Neil's death. How tragic for you and your families! As you have heard a thousand times, the world has lost a humble hero.

Our son, Wes, will never forget the day he spent with Neil at Wright Patterson Air Force Base. (Wes is studying astrophysics in college.) Neil was so kind and generous with his time that day. Wes cherishes the memories. Dan and I never had the privilege of getting to know Neil, but those that did only had wonderful things to say about him.

Death is hard on the ones left behind. We are sorry for your loss and we will keep you and your family in our prayers. And yes, when we look up into the night sky and see the moon, we will wink and think of Neil and his accomplishments for all of mankind.

Sincerely,

Dan and Kellie
Dan and Kellie Peters
Cincinnati, OH

Daniel S. Peters (b. 1950) is president of the Lovett & Ruth Peters Foundation in Cincinnati. The focus of the foundation is on improving the quality of K–12 education nationwide. The foundation was established in 1994 by Dan's father, the oil and gas entrepreneur Lovett C. Peters (1913–2002), and mother, Ruth Peters (1917–2009). For seventeen years Dan worked at Procter & Gamble in a variety of executive positions in advertising, purchasing, and R&D. He is the former chair of the Philanthropy Roundtable, a national association of grant-makers. He is also cofounder of the Alliance for Charitable Reform, a Washington, D.C.–based organization advocating "common sense reform" of the nonprofit sector, and serves on several boards, including Hillsdale College and Catholic Education Partners. But his main work has been carrying on the legacy of his parents' foundation. In recent years, the Peters Foundation has focused its giving on technology-enabled reforms like blended learning, which enables schools to serve more students more cost-effectively. For example, the foundation has funded the expansion of the Carpe Diem e-Learning Community to Cincinnati and Indianapolis.

"NEIL WAS A GIFT TO THIS WORLD"

September 1, 2012

Dear Carol,

My heart goes out to you and your family.

Neil was a gift to this world and I feel honored to have known him . . . such a humble, dynamic and genuine man.

My father-in-law collected stamps and I found these. I wanted to pass them on to your family.

With love,

Ginna Portman Amis
Minneapolis, MN

In this letter to Carol, Ms. Amis sent a full pane of USA 18-cent stamps from 1981 illustrating different aspects of the United States' achievements in space exploration.

Neil and Carol were close friends of the Portman family, notably Robert Jones Portman (b. 1955), the U.S. senator (and former congressman) from Ohio, and his wife, Jane, and Rob's sister Ginna Portman Amis and her husband, Allen. The family business, Portman Equipment Co., was founded by their father, William C. Portman (1922–2010), in 1960. The company sold and distributed Clark forklifts, later switching to Caterpillar lifts, while expanding into other product lines, including equipment used to move material in distribution centers, floor cleaning equipment, and power and storage systems. In 2004 the family sold the company to a Dutch conglomerate. Rob Portman was one of the speakers at Neil's private funeral in Cincinnati.

"A MAN OF GREAT QUALITY"

September 1, 2012

Dear Carol,

The news of Neil's sudden death came as a great shock to us.

Only 2 weeks ago I sent him an e-mail, wishing him a full recovery and expressing the hope that I would see him at the A-A Ranch.

He responded immediately and wrote "Attendance is unpredictable, but looks doubtful."

Unfortunately there is no doubt any more. We will never meet again, but the memories remain.

We shared many fine moments and I have lost a dear friend.

But your loss is much greater. He was a man of great quality, much admired by many.

We wish you and your family lots of strength in these trying days.

Our thoughts and prayers are with you all. He will never be forgotten.

With our deepest sympathy,

Karel
K. en M. Ledeboer
Heemstede, The Netherlands

A RELUCTANT ICON: LETTERS TO NEIL ARMSTRONG

Karel Ledeboer (b. 1937) is a retired Dutch aircraft engineer and global aviation expert who has held leading executive positions with KLM Royal Dutch Airlines, the International Air Transport Association (IATA), Swiss International Airlines, 2nd Opinion Aviation Management Services, and the Flight Safety Foundation in Washington, D.C. A 1962 graduate of the Technical University in Delft, he is a fellow of the Royal Aeronautical Society and has received a knighthood in the Order of the Netherlands Lion.

"LOSS OF THIS GREAT, SELF-EFFACING, MODEST AMERICAN HERO"

September 2, 2012

Dear Carol,

We were very saddened to hear that Neil's surgery wasn't going well and that he didn't pull through. Our prayers go out to you and the family over the loss of this great, self-effacing, modest American hero. The many articles we've read over the past couple weeks express his greatness in a very memorable manner.

Paula and I are counting on seeing you in Boca Grande this winter.

Love,

John
John S. Lillard
Lake Forest, IL

John S. Lillard (b. 1930) is the former chair of the Wintrust Financial Corporation. Along with his wife, Paula—who wrote a separate letter of condolence to Carol Armstrong dated August 26, 2012, included earlier in this chapter—John Lillard has been highly active in a number of educational, environmental, and philanthropic causes, notably through the Red Bird Hollow Foundation, a nonprofit founded in Lake Forest, Illinois, in 1991. Upon learning that his letter was being published in this book, Mr. Lillard wrote to me: "It is wonderful that you are helping us all know, remember and understand this incredible man, and all he has given to our earth and humanity."

LETTERS FROM A GRIEVING WORLD

"TRYING TO STAY BELOW THE PUBLIC'S RADAR"

September 2, 2012

Dear Carol,

Please accept my sincerest sympathy on Neil's passing. You are in my thoughts and prayers. I know there are no words that you have not already heard, but know you are in the hearts of your many friends.

There are so many memories for Neil and George. I think of the stories George would tell about HARP and about travelling with Neil and trying to stay below the public's radar. They had a great friendship and respect for each other.

I wish you peace, Carol.

Fondly,

Ellen
Ellen Rieveschl
Covington, KY

In the envelope with the card to Neil, Ms. Rieveschl put a clipping of a recent Reuter news story published in the Chicago Tribune, *headlined "Some See Stars, Dollar Signs in Signed Items/Value of Armstrong Memorabilia Skyrockets." The story began, "A postcard signed by the first man on the moon, Neil Armstrong, has sold for $2,384, more than three times its presale estimate, signifying a 'hot market' since the astronaut died." Ms. Rieveschl underlined that sentence and in the margin of the clipping, she wrote to Carol, "Not what he would have wanted."*

Ellen F. Rieveschl (b. 1945) was the wife of the noted chemist George Rieveschl (1916–2007), the inventor of the popular antihistamine Benadryl. A 1933 graduate of the Ohio Mechanics Institute of Technology, George Rieveschl looked during the Depression era for a job in commercial art. Failing that, he enrolled at the University of Cincinnati—paying a tuition of $35 per semester—earning a bachelor's degree in 1937, a master's in 1939, and a doctorate in 1940. During World War II, he went to work at Parke-Davis, where he eventually rose to vice president of commercial development. In 1970, the same year Neil Armstrong accepted a faculty position at the University of Cincinnati, Rieveschl returned to his alma mater as vice president of research. In 1975 he joined with Armstrong, along with Dr. Henry Heimlich, Cincinnati's famous inventor of the Heimlich maneuver who was practicing medicine at Jewish Hospital, and Edward A.

Patrick, a professor of electrical engineering, to create a campus research institute they called HARP (short for Heimlich-Armstrong-Rieveschl-Patrick). The HARP Group set out to use technology affiliated with NASA's Apollo program to design an artificial oxygenator for the treatment of chronic lung diseases. The result of HARP's efforts was the Apollo double diaphragm pump, a modified version of a pump used by NASA to circulate temperature-regulating fluid in its space suits. Although the pump was never completed, the research conducted by the HARP Group led to the design, patent, and manufacture of a successful device called the Heimlich Micro-Trach, an oxygen delivery system that allowed patients to use much less oxygen than they would otherwise need by delivering the gas directly to the windpipe, where it could not escape in breath as easily.

Dr. George Rieveschl retired from the University of Cincinnati in 1982. Five years later, the university named its new science and engineering building in his honor.

"THE BRILLIANT, INQUISITIVE, COURAGEOUS FIRST MAN"

September 3, 2012

Dear Carol,

The proceedings on Friday were powerful, uplifting and right on target—perfectly depicting the two elements of Neil Armstrong: the brilliant, inquisitive, courageous First Man on the Moon and wonderful, unassuming man whom we loved and admired and cherished as a friend. It has been an extraordinary experience for us.

We are aware of the pain which his loss has inflicted on you and your family. Neil is physically gone, but you are here, with whom we will share the memories of an American hero and a wonderful friend. You are in our thoughts and prayers.

With our love,

Joe
Joseph H. Head, Jr.
Louise Head
Cincinnati, OH

A graduate of The Taft School, Yale University, and Harvard Law School, Joe Head (1932–2019) had a distinguished career as legal counselor for Graydon Head & Ritchey, an esteemed law firm established in downtown Cincinnati in 1871, which in the ensuing decades expanded into Northern Kentucky and Southeastern Indiana. Through six decades of active civic engagements, Head became one of Cincinnati's best-known citizens. He served as trustee and president of the Children's Home of Greater Cincinnati and sat on the boards of the Fifth Third Bank and Christ Hospital. He also served as chair for the Cincinnati Fine Arts Fund campaign, Hamilton County Republican Party, Greater Cincinnati Chamber of Commerce, and Cincinnati Business Committee. Head was married to Louise Atkins Head for sixty years. Prior to attending Harvard, he served for two years in the U.S. Army's artillery division in Germany.

"CAUGHT THE ESSENCE AND SPIRIT OF NEIL"

September 3, 2012

Dear Carol,

Friday was a fitting tribute for Neil. All four speakers were appropriate and certainly caught the essence and spirit of Neil. Charlie was particularly effective, being in line with a celebration of a close friend's life.

My remarks at the Goodman Tournament seemed well received and I hope you approve.

Of course, many people will miss Neil, none more than you.

Connie joins me in sending our deepest sympathies to you.

Tuck
Taylor Asbury
Cincinnati, Ohio

Dr. Taylor "Tuck" Asbury (b. 1921) is a retired Cincinnati ophthalmologist who became one of Neil's good friends later in his life. "Charlie" refers to Charles S. Mechem Jr., the former president of Cincinnati-based Taft Broadcasting, who became Neil's close friend after Neil came to serve on Taft's board of directors. Mechem gave one of the eulogies at Neil's private funeral held at the Camargo Club, of which Tuck Asbury was a member. The following are remarks that Dr. Asbury presented at a wake held for Neil by Camargo Club members on September 1, 2012, the day after the private funeral.

A RELUCTANT ICON: LETTERS TO NEIL ARMSTRONG

Remarks by Taylor Asbury at the Camargo Club
Cincinnati, Ohio
"Neil Armstrong—Everybody's Hero"

September 1, 2012

We are very fortunate that Neil had the good judgment to choose Cincinnati to live in the last part of his life. The Camargo Club is particularly fortunate in having him as a member which allowed many of us to get to know this remarkable man on a personal basis. Yesterday many of us heard four excellent speakers paint a wonderful picture of Neil's life and accomplishments.

Everyone knows of his remarkable career and eternal fame as the First Man to step on the Moon as well as his eternal words upon doing so. "That's one small step for man, one giant leap for mankind." These are the words of a humble man, not scripted by someone else.

There has been a widely held impression that Neil was reclusive, but this is far from true. This impression is derived from his unwavering intention not to use his fame for personal or financial gain, which validates NASA's good judgment in selecting him to be the First Man. While he never initiated public appearances, when called upon, he often spoke in appropriate settings.

For those who knew him well, he was always upbeat, outgoing often humorous and always fun to be with. About a dozen of us socialized with him six or eight times a year at the informal Rookwood Historical and Philosophical Society, the meetings of which included cocktails and dinner, followed by a low stakes poker game which Neil never missed when in town. In this setting we knew the real Neil.

Neil gave of himself after his astronaut days. For example, he headed a successful $3.5 million fund raiser for his alma-mater, Purdue. He helped promote many other worthwhile fund raisers. Occasionally he did comment on some current government policy, such as being critical of the recent downgrading of the space program and reduction of the military, but he generally kept politics out of his public statements. For anyone who wishes to know more about his professional and private life, his only authorized biography entitled First Man, written by James Hansen and published by Simon and Schuster in 2005 is highly recommended. Neil told me that he read the book before publication and suggested no changes, even though he did not agree with a few of the author's interpretations of his life. Typical Neil.

Today we played the Sixty-Fifth Annual Men's Foursome for the Timothy S. Goodman Trophy, a tournament Neil particularly enjoyed. Becoming a Camargo Club member in 1995, he played in the tournament in 1996 which was the first year that included a Senior Division. He was on the winning team helped by teammates Bob Gerwin, Guy Randolph, and Bob Wersel. Neil played in all but one Goodman thereafter, being on the winning team five more times. His six wins is the tournament record for Seniors. I was lucky enough to partner with him 14 of the last 15 years, and he had intended to play again today. Scott Brinker and I helped him in four of his victories. Neil must have been looking down on us today, as one could not help but feel his presence. Perhaps it would be appropriate to name the Senior Division the Neil Armstrong Division of the Goodman Tournament.

I knew Neil liked golf, but found out yesterday from Charlie Mechem, and Neil's sons, Rick and Mark, that golf was a real passion of his.

Many celebrities do not live up to their reputation but Neil is a notable exception. Thanks for the memories Neil.

Surely God blessed him.

"AN INSPIRATION TO ME, AS A GENTLEMAN"

September 4, 2012

Dear Carol:

My thoughts and prayers go out to you during these difficult times. Jo and I have been thinking & reminiscing of the times we had together with you and Neil in Denmark & in Tulsa. We talked about how you wanted to come by & see our kids! Those were great times! Thank you for being such a great friend to my mother & father. Hopefully I can repay you in kind in the years to come.

I will miss Neil at the Ranch! He was a "guy's guy" in that environment and was always so pleasant to be around. He is an inspiration to me, as a gentleman. So, thank you for taking care of my friend Neil. I hope you find some peace & strength during these times. I miss you & will keep you in my thoughts. Jo sends her warmest blessings too.

Love,

Tray
Colonel Tray Siegfried
Tulsa, OK

Raymond H. "Tray" Siegfried III (b. 1943) is today the sole owner of Vertical Aerospace, an airplane maintenance, repair, and overhaul service provider located in Bristow, Oklahoma, southwest of Tulsa. A veteran pilot and commander of the 138th Fighter Wing of the Oklahoma National Guard, Siegfried participated in numerous overseas operational deployments, including Operation Iraqi Freedom, where he flew numerous combat missions in the F-16 fighter jet. A holder of an MBA from the University of Tulsa and an undergraduate degree in business from the University of Notre Dame, Colonel Siegfried served in various capacities in aerospace industry trade associations, including the General Aviation Manufacturers Association and Aerospace Industries Association. His wife, Jo, is a graduate of Dartmouth College and a former Wall Street investment banker and financial analyst. Siegfried's late father, Raymond H. Siegfried II (d. 2005), was the founder of The NORDAM Group, which from its birth in 1969 as a local Tulsa manufacturing company grew into a global aerospace corporation operating on four continents. During its most rapid expansion, Tray worked for NORDAM as its vice president of sales and strategic resources.

"HARD TO BELIEVE THAT NEIL IS GONE"

September 4, 2012

Carol,

Again I want to express my deep sympathy to you and the family. It is hard to believe that Neil is gone. He was a dear friend and I had the honor to work with him during our tour with the astronaut group.

In the mid 1960s Neil and I were assigned to work on and evaluate the flight control system for the Lunar Module which was located at the simulation facility at North American Aviation in Columbus. We made trips there together and on several occasions I stayed at his parents' house. After I flew Gemini 6, Neil commanded Gemini 8 and I commanded Gemini 9. I was the backup commander for the 2nd Apollo flight. Then I commanded Apollo 10 to the Moon, and he commanded Apollo 11.

It was good to help supply a lot of the facts for this letter and testimony where Neil, Gene and Jim were the major turning point to turn around the Obama Administration's position on space exploration. Without Neil's position as leader of the group it would be impossible to achieve this goal.

Linda's brother has suffered from terminal cancer and the family called Tuesday and requested that she come to Oklahoma as soon as possible since he would die within a week to ten days. She left yesterday and I am here with our Russian son Stas, who is a senior in high school. I need to be here with him and will be unable to attend the national memorial in Washington D.C.

I regret that I will not be able to be there, but Neil will be in my thoughts and prayers.

God bless.

Sincerely,

Tom Stafford
Cape Canaveral, FL

Thomas P. Stafford (b. 1930) joined NASA, as Neil Armstrong did, in September 1962 as part of the second group of astronauts. A 1952 graduate of the U.S. Naval Academy, Stafford entered military service as a second lieutenant with the U.S. Air Force. Following pilot training, he was assigned in 1954 to the 54th Fighter-Interceptor Squadron at Ellsworth AFB, where he flew the F-86 in Arctic defense; the next year he transferred to the 496th Fighter-Interceptor Squadron in West Germany. In 1958 Stafford attended the Air Force Test Pilot School at Edwards AFB, finishing first in his class; after graduation he remained at Edwards as a flight instructor. With NASA in 1966 he flew America's first rendezvous in space on Gemini 6 and later that year piloted Gemini 9's orbital rendezvous mission. In May 1969 he commanded Apollo 10, the first flight of the lunar module to the Moon, in what amounted to a dress rehearsal for the first Moon landing a few months later by Apollo 11. In 1973 he commanded the Apollo-Soyuz Test Project (ASTP) spacecraft that docked with a Soviet Soyuz spacecraft in Earth orbit as a symbol of détente between the U.S. and USSR. In the late 1980s Lt. General Stafford (promoted to that rank in 1972) chaired a team to advise NASA on how to carry out President George H. W. Bush's space policy, completing a study entitled "America at the Threshold." He also cochaired the task force that in 2003–2004 assessed NASA's implementation of the recommendations made by the Columbia Accident Investigation Board for the space shuttle's return to flight. In 2019 Stafford received the General James E. Hill Lifetime Space Achievement Award from the Space Foundation, one of many prestigious awards he has received over his illustrious career. In his hometown of Weatherford, Oklahoma, the Stafford Air and Space Museum, a Smithsonian affiliate, opened to the public in June 2010.

Stafford and his wife, Linda Ann Dishman—whom he married in the late 1980s following three previous marriages—adopted two sons, Michael Thomas and Stanislav "Stas" Patten.

"WE'LL NEVER LOOK AT THE MOON IN QUITE THE SAME WAY AGAIN"

September 4, 2012

Dear Carol,

We send our deepest condolences to you and your family. It was a joy to know Neil from both Purdue and Telluride.

What fond memories we have of you and Neil in our home for dinner. Remember Ron asking everyone at the table what they were doing when Neil walked on the moon? Neil's response: "I was working." How typical of his unassuming, droll wit.

The many media tributes we've seen about "The Right Man" and the "Reluctant Hero" certainly echo our own feelings. We'll never look at the moon in quite the same way again, and, rest assured, Neil will get many winks from us.

We hope to see you again whenever you visit Telluride.

Sadly, Fondly,

Nancy and Don Orr
Telluride, CO

LETTERS FROM A GRIEVING WORLD

Neil Armstrong and Don Orr (b. 1940) became good friends through their connections to their alma mater, Purdue University, from which Orr graduated with a degree in engineering in 1961. They strengthened their friendship by both owning vacation homes and spending time together with their wives in Telluride, Colorado. Just as Neil cochaired two of the university's major fundraising efforts, Orr was a long-standing volunteer for Purdue's School of Chemical Engineering, heading up a capital campaign that raised $25 million for new construction and renovations. Orr—who was named a Distinguished Engineering Alumnus in 1989—was also a member of the Dean's Advisory Council for the School of Management and a member of the President's Council. Neil and Orr also shared an enthusiasm for Boilermaker sports. The two men, both from Ohio (Orr from Wooster) also shared military experience (Orr was a lieutenant in the U.S. Army) and a career in business (Orr retired as senior corporate vice president from Air Products & Chemicals, Inc.).

"HIS WIDE AND WARM SMILE"

September 4, 2012

Dear Carol,

I was just sitting here thinking about Neil three weeks ago tonight, eating pizza and talking about going to the Paris Air Show next June together, followed by Ireland, where he apparently hadn't played golf but had always wanted to. He told me about possibly traveling to Toulouse, at Tom Ender's invitation, to come fly the A380. He was more fun. "Jose," he'd say when greeting me, with his wide and warm smile. He was definitely a cute guy, whom all will miss greatly. None more than I.

Your pretty Service was exactly what Neil would have chosen. It was so much him in every respect: colorful United States Honor Guard, beautifully sung September Song solo, famous hymns, spectacular Readers, memorable eulogies, (some verbose), Metropolitan Opera brass, marching Bag Pipers, Cincinnati Symphony violins, and four young aviators in formation, then one Missing Man abruptly pulling straight up into the heavens. Meanwhile, soaring through outer space, Neil is still applauding you loudly, Carol, as he always has.

If there was a kind of consolation to this sad time, for me, it was your knockout, precious children and adorable grandchildren, your lovely friends Beth and Gary, your very cool brother, and absorbing, best I could, your unforgettable lesson in courage, grace, and poise.

I have the picture of you and Neil that I framed last Christmas morning, just before you arrived. I remember your recognizing it as the previous year's Christmas card, but I'm telling everyone it's an original photograph.

Don't hesitate to email me if anything simple comes up; or text me if you're out of peaches or wine. Really, I can get there in three minutes if I cut across the golf course.

Faith, Hope, Love abides,

Joe
Joseph W. Hagin, Sr.
Cincinnati, Ohio

Joseph W. Hagin Sr. (b. 1926) was a prominent businessman and philanthropist in northern Kentucky and the Greater Cincinnati area. A resident of the Indian Hill community, as were Carol and Neil Armstrong, Hagin and his late wife, Hannah, served for many years as trustees for the Cincinnati Country Day School, a prestigious coeducational (admitting girls since 1972) college preparatory private school near Indian Hill that was founded in 1926. Their two sons, Joseph Whitehouse Hagin II (ironically his actual middle name considering his future in politics) and Hunt Hagin, both attended CCDS, as did future U.S. senator from Ohio Rob Portman and future Ohio governor Bob Taft (Hagin II and Portman remain close friends to this day). Joe Hagin II (b. 1956) rose to political prominence as a Washington insider and political aide in the White House for President George W. Bush and President Donald Trump. (Hagin II served as Trump's deputy chief of staff for operations into July 2018. During the Obama years he founded the Command Consulting Group, a company that marketed itself as "a global security and intelligence consulting firm that provides advisory services to governments, corporations, and high net worth individuals." In the Trump Administration, he played a central role in planning the North Korea–United States summit held in Singapore in June 2018.) In addition to his political career, Hagin II also has considerable corporate experience. In the mid-1980s he worked as public affairs director for Federated Department Stores, which owns Macy's and Bloomingdale's; in the early 1990s he was vice president of corporate affairs at Chiquita Brands International; 2008–2009 he served as interim CEO of Jet Support Services, Inc., one of the world's biggest providers of maintenance work for business aircraft. After Hagin II left the Trump White House in July 2018, President Trump named him to the new National Commission on Military Safety that was tasked to determine the cause of the alarming rise in military jet accidents.

"NEIL WAS THE BEST AT ALL THREE LEVELS"

September 4, 2012

Dear Carol,

A multitude of people share your grief.

At Eaton it was ideas, education and results. Neil was the best at all three levels. The two of us did not have to discuss these issues. We immediately understood the values and the risks of some of the procedures.

Eaton turned the corner early and became the shining star on the list of companies that stood steady in a mixed-up financial world. But more than that Neil was an example to our way of thinking that led us into not being scared of uncertainties.

Eaton continues to be a first-rate company. Your husband (and my best friend) deserves credit for aiming in the right direction and establishing the principles that sustained the course.

If there are ways I can be of value to you and the family, please call.

Sincerely,

H. G. Pattillo
Decatur, GA

H. G. "Pat" Pattillo (b. 1926) and Neil became very close friends. They talked on the phone frequently and took their memberships on the Eaton Corporation's board of directors very seriously. According to a message to me from Mr. Pattillo, they "also took their card games at Eaton House before each board meeting just as seriously," though they could never remember "who owed who money." In 1950 Pattillo's father, H. A. Pattillo, founded Pattillo Construction Company, Inc., a design-build general contractor based in Clarkston, Georgia (20 minutes northeast of downtown Atlanta), with his two sons, Pat and Dan. Early Pattillo projects included numerous churches, schools, and other commercial buildings before the company began focusing primarily on industrial construction. In a twelve-year period during the 1960s and 1970s, Pattillo constructed 144 industrial buildings in one master planned park in Stone Mountain, Georgia. The company currently thrives as a women-owned general contractor under the leadership of a fourth generation of Pattillos, Bree Pattillo.

"HE SAID LITTLE, BUT EVERY WORD MATTERED"

September 4, 2012

Dear Carol,

I was honored to serve with Neil for a short time on the Eaton Board before he retired, and I only met you briefly at his retirement event.

In that short time, I was able to observe many of the qualities that made him an extraordinary human being. He said little, but every word mattered. He recognized that great boards of directors are team and collaborative endeavors, not showcases for individual brilliance. He was always looking ahead, not behind.

Like many, I was touched in a very permanent and good way by his presence and his leadership.

While we all move on with our lives, I want you to know that Neil made a big difference in my life.

Sincerely,

Mike Critelli
Darien, CT

Michael J. Critelli (b. 1948) rose through the ranks at Pitney Bowes—a global technology company (founded in 1920) that crafts innovative products and solutions in the areas of customer information management, location intelligence, customer engagement, shipping and mailing, and global ecommerce—to become its executive chair and director. Graduating from the University of Wisconsin in 1970 and Harvard Law School in 1974, he joined Pitney Bowes in 1979 after serving as an associate in two Chicago law firms. At Pitney Bowes, he became the company's general counsel in 1988, with the added responsibility in 1990 as its chief of human resources. Critelli became the company's CEO in 1997 and remained in that position until 2008, leading the company through a period in which it and the entire mailing industry underwent transformational change into the world of digital networks. After retirement from Pitney Bowes, he became the president and CEO of the Dossia Service Corporation, a software and technology firm for consumer-driven healthcare that he cofounded with the head of the Intel Corporation and senior leaders of eight other Fortune 500 companies. In 2013 Critelli moved from the board room to the sound stage as the producer of the full-length feature film From the Rough, *the true story of the first African American woman to coach a men's college athletic team (golf), starring Taraji P. Henson in the role of Dr. Catana Starks. Critelli and Neil Armstrong got to know each other well as fellow members of the board of Eaton Corporation. Critelli served on a number of boards, including that of ProHealth Physicians, Inc., CVC Capital Partners, RAND Health Care, Yale School of Public Health, and Regional Plan Association, the latter being a tristate organization whose mission is to improve the New York metropolitan region's economic health, environmental sustainability, and quality of life.*

"HE NEVER CHANGED"

September 4, 2012

Dear Carol:

Always the quiet humble man. He did so much for mankind. Over all the years we have known Neil, he never changed.

We have lost a special friend and words cannot express the sorrow we feel.

Our prayers are for you and the family. Hoping to ease your grief.

Our best to you.

T. R. and Bernardine Swartz
Navy friends and Golden Eagles
Theodore R. Swartz
Poway, CA

Theodore R. Swartz (b. 1935) is one of the great American naval aviators—the recipient of the Silver Star, six Distinguished Flying Crosses, seven Individual Air Medals, twenty-seven Strike Flight Awards, and eight Navy Commendation Medals over the course of a twenty-four-year career (1954–1977). He served two tours in the Vietnam War, flying over 300 combat missions. On May 1, 1967, as a member of the VA-76 attack squadron (nicknamed the Spirits from its motto, the Fighting Spirits of 76), flying the A-4 Skyhawk, Lieutenant Commander Swartz shot down a MiG-17 with air-to-ground rockets, the first and only MiG to be downed by an A-4 Skyhawk during the war. Retiring from the military in 1978 after twenty-five years in the service, T.R. went to work as a defense contractor with the U.S. and various allied foreign tactical aviation forces, providing systems for real time and postflight air/ground combat training and damage assessment.

T.R.'s wife, Bernardine (b. 1934), died on August 5, 2019, what would have been Neil's eighty-ninth birthday.

"HIS WILLINGNESS TO STEP UP"

September 5, 2012

Dear Carol,

I was so sad to see that one of our nation's most accomplished explorers was gone. He handled his historic achievement and the accompanying fame with such humility and grace.

I appreciated his willingness to step up, and be public in the warnings about NASA's direction. His willingness to do so did help turn the direction back the right way.

My thoughts are with you and your family.

Kay Bailey Hutchison
United States Senator
Dallas, TX

Kay Bailey Hutchison (b. 1943) is a former U.S. senator (R-Texas) who currently serves as the U.S. Ambassador to the North Atlantic Treaty Organization (NATO). A graduate of the University of Texas who earned a law degree at the University of Texas School of Law, Hutchison began her career as an attorney in the Dallas law firm of Bracewell, LLP. She subsequently served in the Texas House of Representatives, as the Texas state treasurer, and as vice chair of the U.S. National Transportation Safety Board. Her time in the U.S. Senate came in a nonpartisan special election in 1993, in which she defeated the Democratic incumbent and became the first female senator in Texas history.

Hutchison won reelection to the Senate in 1994, 2000, and 2006. In 2010 she ran unsuccessfully to be the Republican candidate for governor of Texas, losing the primary to incumbent Rick Perry. When she left the Senate in 2013, she was the most senior female Republican senator and the fifth most senior female senator overall. In 2013 she joined the law firm of Bracewell & Giuliani. Hutchison attended Neil's private funeral at the Camargo Club. A few days later, from the Senate floor, she honored Armstrong and the legacy of Apollo 11 and warned against cutting items in the space program, which she believed could ensure a healthy national economy in the future. On September 1, 2012, in her weekly column, Notes from the Capitol, she published her tribute, "Neil Armstrong: A True American Hero."

"THE BEST KIND OF EXAMPLE"

September 6, 2012

Dear Carol:

Ellen and I extend our deepest sympathies on the passing of Neil. I needn't tell you that he was "one of a kind." The manner in which he conducted himself was the best kind of example he could have provided others who follow. As you know, it was my great pleasure to serve on the United Airlines Board and to sit next to Neil for eleven years. I was a large beneficiary of his friendship as was the airline because of his keen intellect and total commitment.

We so much enjoyed the opportunity to spend time with you both at Lake Geneva a few years ago. My only regret is that we did not have more time to do this more frequently.

You are in our thoughts and prayers.

Sincerely,

Jim
James J. O'Connor, Sr.
Chicago, IL

A 1958 graduate of the College of the Holy Cross, a Jesuit University in Worcester, Massachusetts, James J. O'Connor Sr. (b. 1936) is the retired chair and CEO of Commonwealth Edison, Illinois's largest utility company, and cofounder and cochair of the Big Shoulders Fund, a charity that supports Catholic education in Chicago (O'Connor grew up in Chicago's South Side). He is the former director and chair of Armstrong World Industries and, for over twenty-eight years, a director of United Airlines. Civically active, O'Connor provided his leadership to a number of Chicago's cultural institutions and nonprofits, including the Field Museum of Natural History, the Museum of Science and Industry, the Chicago Symphony, Catholic Charities, American Cancer Society, and United Way. His son, James J. O'Connor Jr., is the managing director of William Blair & Company, a global investment banking and wealth management firm.

"THE CHRISTIAN VIRTUE OF HUMILITY"

September 7, 2012

Dear Ms. Armstrong,

I am sorry for your loss. Not only did I have the deepest admiration for your husband for the Gemini-Titan 8 mission and Apollo 11, but he displayed the Christian virtue of humility like no one possibly since Jesus Christ.

I know what you're going through, after 31 years of marriage I lost my best friend, my late wife Joan in November of 2009. It happened so suddenly I didn't get to say good-bye to her. I'm 63 years old, a retired steelworker and I was paralyzed in a car accident in 1999. Don't feel sad for me, there are far more people worse off than I. If you ever feel that you need someone to talk to, I'm at [email address withheld] or [phone number withheld]. Just remember they are where we all strive to be.

Sincerely,

Tom Zaborski
Las Vegas, NV

Thomas Steven Zaborski (b. 1949) is a retired steelworker for the Inland Steel Company, an integrated steel company with a large mill on Lake Michigan in East Chicago, Indiana, that operated from its establishment in 1893 to 1998, when the business was acquired by Ispat International N.V., a steel producing company with operations in Mexico, Trinidad, Canada, Germany, and the U.S.

LETTERS FROM A GRIEVING WORLD

"THE LITTLE BOY WHO RAN UP TO MR. ARMSTRONG"

September 6, 2012

Dear Mrs. Armstrong,

I wish to express my sympathy for your loss. Mr. Armstrong was my hero. I was born the same year of Apollo XI's 30th anniversary. I hope someday to take another giant leap for mankind, on the moon or Mars. Space flight just seems amazing to me, just being able to see everywhere you've ever been or wanted to be, in just a glance.

I don't know if you remember, but I was the little boy who ran up to Mr. Armstrong and shook his hand while you were leaving the capitol after the congressional gold medal ceremony last November. I'm pretty sure you were right next to him.

The influence Mr. Armstrong has been in my life so far, I know I'm in the right direction to complete my hopes and dreams. I know he was a great man, even though I was in his presence for no more than fifteen seconds, I know he was a great man. I'm sorry, and I'm sure me and every other American will miss him too.

Sincerely,

James Kelly
Lexington Park, MD

On the back of the envelope, young Mr. Kelly printed: "From: 12 YO Boy, Please read, the sender shook hands Mr. Armstrong Nov 16, 2011, U.S. Capitol."

Born in 1999, James D. Kelly Jr. was twelve years old when he attended the congressional gold medal ceremony for the Apollo 11 crew at the Capitol Building in November 2011. A boy who dreamed of one day being an astronaut, J. D. wrote letters to a number of astronauts, asking for information about how to become an astronaut, receiving responses from several of them but, regrettably, not from Neil. Today, J. D. is twenty years old and a student at Stanford University studying electrical engineering; he is also a U.S. Air Force ROTC cadet. His grandfather, Keith Kelly, was a chemical engineer who worked for NASA at Cape Canaveral in the 1960s. J. D.'s father, James Thomas Kelly, went to elementary school near his home in Cocoa Beach, Florida, and remembers his class going outside to watch many of the Mercury, Gemini, and Apollo launches. After graduation, J. D. hopes to be commissioned in the Air Force.

"CHERISH HIS FRIENDSHIP"

September 7, 2012

Dear Carol,

I missed Neil very much at the Conquistadores meeting, from which I returned yesterday afternoon and, believe me, I was not alone in this respect. Jim McDivitt made some very informative and touching remarks about Neil, which were affectionately received.

Today, I found the invitation to the Memorial Service on Thursday, and I am profoundly grateful to you for inviting me to attend. Unhappily, when I got back I found that I was scheduled for some surgery on Thursday (*not* life threatening) so I will not be in personal attendance but will be with you in my heart and thinking of Neil, not just Thursday but every week for the rest of my life.

I once sent Neil a note telling him that I would cherish his friendship and look up to him as a role model even if he'd landed in Los Angeles instead of on the moon. I meant that sentiment with my entire heart.

I join you in your grief and will be winking at the moon with you.

Sadly,

Herb
Herb Kelleher
Dallas, TX

Herbert D. Kelleher (1931–2019) was the cofounder and later CEO of Southwest Airlines. A graduate of Wesleyan University with a law degree from New York University, Kelleher moved to Texas with the idea of starting a law firm but instead, with Dallas banker John Parker, came up with the concept of a no-frills, low-cost, intrastate airline serving the Texas Triangle (Houston, Dallas–Fort Worth, San Antonio, Austin). From this start in 1971, Southwest Airlines rapidly crew to become the world's largest low-cost carrier with total assets of over $26 billion. Over his distinguished career, Kelleher received over one hundred awards and honors in the worlds of business and aviation, including his induction in 2008 into the National Aviation Hall of Fame. Like Neil, he was a member (from 1990) of the Conquistadores del Cielo, a private club of high-level airline and aerospace industry executives. He received its highest honor, the Big Horse Award, in 2005.

LETTERS FROM A GRIEVING WORLD

"HE EXEMPLIFIED THE BEST IN THE ASTRONAUT OFFICE"

September 7, 2012

Dear Carol,

The Memorial service to Neil was beautiful. He exemplified the best in the astronaut office. I'm so glad we could be there to give you and your family hugs. Thank you for all you did to take care of us so graciously.

Please know that we'll continue to send our love and prayers to you through your great loss.

God bless you.

Dotty & Charlie Duke

P.S. We'll be out of the country during the time of the memorial service in Washington. May the Lord be with you & comfort you.

Mr. and Mrs. Charles M. Duke
New Braunfels, TX

Charles M. Duke Jr. is a former American astronaut who walked on the Moon in April 1972 as the lunar module pilot for Apollo 16 (commanded by John Young). Duke is also well known for his role in the first Moon landing, Apollo 11, for which he served as the capsule communicator (CapCom) in direct voice contact with Armstrong, Aldrin, and Collins from his seat in Mission Control in Houston. Born in 1935 in Charlotte, North Carolina, he graduated from the Admiral Farragut Academy in St. Petersburg, Florida, in 1953, received a BS degree in naval sciences from the U.S. Naval Academy in 1957, earned an MS degree in aeronautics and astronautics from MIT in 1964, and graduated from the Air Force Aerospace Research Pilot School in 1965. Logging 4,147 hours of flying time in various aircraft, including 3,632 hours in jet aircraft, Duke became an astronaut in 1966.

Charlie had an identical twin brother, William (d. 2011), who became a medical doctor. The twin brothers looked so much alike that the appearance of William inside Mission Control during Charlie's Apollo 16 Moon walk shocked everyone present; it was a practical joke that Charlie had planned.

Following retirement from NASA on January 1, 1976, Duke entered the Air Force Reserve and, as a colonel, served as a mobilization augmentee to the Commander, Air Force Basic Military Training Center, as well as the Commander, Air Force Recruiting Service. He also attended the Industrial College of the Armed Forces, graduating in 1978. The next year the air force promoted Duke to brigadier general, a rank in which he stayed active until his retirement from the military in 1986. He stayed active in the Society of Experimental Test Pilots and from 1996 to 2000 served on the NASA Advisory Council. He also became director of the Young Astronaut Council and was on the board of the Astronaut Memorial Foundation, serving as its chair from 2011 to 2018.

Following his military service, Duke became active in business, administering a successful Coors beer distributorship and serving on the board of directors for a number of aerospace-related companies. From the late 1970s on, both he and his wife, Dotty, became very active in different Christian organizations, with Charlie taking leading roles in the Full Gospel Business Men's Fellowship, Christian Business Men's Committee, and International Fellowship of Christian Businessmen. He also has dedicated much of his life to prison ministry.

Charlie Duke has been married to Dorothy "Dotty" Meade Claiborne of Atlanta, Georgia, since 1963; they have two sons, Charles M. Duke III (b. 1965) and Thomas C. Duke (b. 1967), and nine grandchildren. From their home in New Braunfels, Texas, northeast of San Antonio, Charlie and Dotty have traveled together all around the world, with Charlie very popular as a speaker and consultant.

"THE MOST INTEGRITY OF ANY MAN I HAVE EVER KNOWN"

September 7, 2012

Dear Carol:

I thought I would wait to send this because I knew there would be such a flood dumped upon you in the last several days. I am probably being naïve, because I imagine the cards and letters are still swamping you.

What can I say except how sorry I am for you to have lost your mate. The last time I saw Neil was when you and he had Cheryl and me in your home for lunch when the rain canceled our golf plans.

In spite of the large numbers of people after him, Neil always had time to fit me in somehow. I tried to be sensitive about his time and privacy over the years and maybe that is the reason I continued to be part of his life.

When I first met Neil (at the Phi Delt house at Purdue) I was very much in awe of him. He had just returned from the Korean War and had all of the jet fighter experiences to tell us about. I had always wanted to fly myself and his stories were so special to me. He had never been to the 500-Mile race at that time, so I took him to Indianapolis to see it. We stayed together at my home in Indianapolis the night before the race. So, my mom and dad got to meet him too. That was in 1954 I believe.

Another thing that kept us connected was because Dean had been my pledge brother and we knew each other quite well. That ties in to a fun little story that you probably have never heard. Several years ago you and Neil came down to Naples for a Marathon oil meeting. Neil had some free passes to play golf at the fancy Greg Norman golf course, Tiburon. So he called Dean and asked him to get me to play as a threesome there. We did that for the first nine holes. But on the tenth tee box, some guy pulled up by himself and asked if he could join us for the back nine. Since I knew about Neil's desire for privacy, I introduced us by first name only. We all had a great time and shook hands with that guy at the end. To this day, he has no clue that he played nine holes of golf with the first man on the moon.

My memories of Neil will always be of a humble, very bright man possibly with the most integrity of any man I have ever known. I will miss him very much.

God bless you, Carol. I hope your days in the future will be filled with joy after you are able to get past this dreadful nightmare that you must be going through right now. You are in my prayers every morning when I pray for my friends who have lost their mates.

Best Regards,

Joe Fuller
Carmel, Indiana
P.S. No reply necessary, Carol. You have too many.

Joseph Fuller (b. 1935) founded Fuller Engineering in 1966. Located in the Broad Ripple area some six miles north of downtown Indianapolis, the company, which is still in business, is an engineering sales firm specializing in the sale of HVAC, security, and fire suppression products. (When he retired in the early 2000s, he sold the company to five of his employees.) Fuller was a graduate of Purdue University, where he became a good friend of Neil's brother, Dean Armstrong, who is mentioned in the condolence letter. For many years Dean and his family lived in nearby Anderson, Indiana, where he was a foreman and later manager at a Delco-Remy plant that manufactured starters and alternators for General Motors automobiles and trucks.

"HOLY MACKEREL—NEIL ARMSTRONG!"

September 7, 2012

Dear Carol—

I met Neil many years ago when we were both new members of the Question Club. He subsequently resigned his membership because of an annual aerospace meeting in conflict with the Question Club dates.

Shortly after his landing on the Moon, Minor and I bumped into him at Crested Butte, Colorado, skiing at Christmas. We were with a large group of Minor's family and friends from Columbia, Mississippi. I asked Neil if he would join the group for a drink at our Christmas Eve party and he graciously agreed. So for several days I told all these people I was bringing a surprise guest to the party. All the kids thought it would be some friend dressed as Santa Claus. We knocked on the door & a young boy about 9, who was the son of a Reserve Air Force pilot, answered & without hesitation turned to the party inside & yelled out, "Holy Mackerel—Neil Armstrong! That's better than Santa Claus!" We stayed for a drink & Neil was the special feature of the party. Minor & I will never forget how much joy he added to the life of those unsuspecting guests—especially the kids!

We remained friends for many years even though our paths didn't cross often enough.

He will be missed by all his friends for the kind of person he was as much as for his accomplishments.

Our thoughts & prayers are with you and his family.

Sincerely,

Minor & Danny

Daniel W. LeBlond
Cincinnati, OH

Daniel W. LeBlond (1927–2019) presided over R. K. LeBlond Machine Tool Co., founded by his grandfather in 1917. He ultimately guided the company through its acquisition by Makino Milling Machine Co. of Tokyo, Japan, in 1989. A graduate in mechanical engineering from Purdue University, LeBlond served in the U.S. Army following World War II. He was married to the late Minor Morrow LeBlond for sixty-five years, with whom he had two children. LeBlond served as a member, officer, and director of numerous civic and commercial boards of directors in support of the Greater Cincinnati community, as did his wife.

"NEIL WAS JUST A REALLY GOOD GUY"

September 11, 2012

Hi Carol.

First I want to apologize for typing this as I wanted it to be more personal. However, my handwriting is abysmal!

As you may remember, I was a graduate student of Neil's at UC in the early years of his time there. I was the guy who wrote the story for his eightieth about the flight test project we did together. I explained how he went to the effort of getting a plane for us to use and somehow got funds for the work we had to do to complete the project so I could have that experience and write a thesis for graduation. As I mentioned in my birthday story, Neil really went out of his way on that one and I've never forgotten it. I can't say I was a very good student but I can say I learned a lot on that experience and from him. I'll always be thankful.

After school, over the years, we stayed in touch. I really appreciated him and his willingness to keep in contact. Every time I wrote him an email or called he would answer. We had some nice lunches, more recently at Camargo, where we had a good time talking about airplanes and a little about business. He always had good stories and good advice. As an example of a story I really enjoyed, last year when we were together, he told me about his time in the Navy delivery squadron. After Korea, he still had time to serve so he was assigned to move airplanes to and from depot maintenance shops and the like. He got to fly all kinds of Navy airplanes, to the point where his flight suit pockets were full of checklists for the various types he would be assigned to fly. He said the only problem was to make sure he always pulled out the right checklist for the particular plane he was in. I got the impression he really got a kick out of that job. Truly a pilot's dream.

All this to say I really miss him. I know there's a lot of talk about how private he was and how he might have been a little mysterious. But, as my friend Dave Parlin said recently, It's very simple: Neil was just a really good guy.

As will you and many the world over, I will really miss him.

Nick Campbell
Denver, Colorado

As stated in his letter, Nicholas Campbell was a graduate student who took classes with Neil Armstrong at the University of Cincinnati in the early 1970s, soon after Neil started teaching there.

"TALKED SO HIGHLY AND FONDLY OF NEIL'S KINDNESS"

September 11, 2012

Dear Carol,

I am so very sorry for your loss of your husband, Neil. You don't really know me, but Dave and I are members of Camargo Country Club and your "neighbors" on Brill Road. We were so sorry we couldn't attend Neil's funeral. My family has known Neil for a long time. My father, Cal Slattery, is from Wapakoneta and graduated high school with Neil. Neil was so wonderful to attend so many high school reunions in past years. My father and my late mother talked so highly and fondly of Neil's kindness. He was a remarkable, humble, and gracious man! My father is so proud of a photo of a class reunion where he is standing next to Neil. He shows it to everyone who comes into his room at the assisted living facility where he lives in New Bremen, Ohio. Our thoughts and prayers are with you at this sad time.

In deepest sympathy,

Kim Dougherty
Cincinnati, Ohio

Kim Dougherty has worked for many years as an exercise specialist at Jewish Hospital in Cincinnati. Her husband, David Dougherty (b. 1956), is a Procter & Gamble alumni and CEO of Education at Work, a nonprofit he founded in 2012. He is also the former CEO of Convergys Corporation, a call center operation. In 2018 he launched a tennis apparel company even as he competed on the USTA Pro Circuit in the Men's 60–65 division. In 2016 Kim and Dave were honored by the Southwest Oho Chapter of the Juvenile Diabetes Research Foundation as "Cincinnatians of the Year." Dave currently serves on the JDRF board and led the 2010 Greater Cincinnati United Way campaign. The JDRF, founded in New York City in 1970, is the world's leading funder of type 1 diabetes research.

"FITTING TRIBUTE TO AN EXTRAORDINARY MAN"

September 12, 2012

Dear Carol,

I shall remember forever Neil's memorial at Camargo. Every detail was a fitting tribute to an extraordinary man. Each speaker confirmed his genius, his bravery, his humor and humility. But what I remember is dinner with you two and the Abbotts at the Quarter and what fun we had. My last happy memory was sitting next to Neil on Dave Hall's hearth as we sang our hearts out and listened to wonderful jazz.

I have enclosed an article from our Vero Beach [Florida] weekly newspaper, the title of which is our zip code. Perhaps you know this story, but if you don't, I thought you would enjoy it.

Saint and I send our love and our prayers that the days ahead will grow less painful with time and the memories of Neil will sustain you. How blessed he was to have you and how blessed you were to have had the love of such a man.

Gay and St. John Bain
Mr. and Mrs. St. John Bain
Cincinnati, OH

A RELUCTANT ICON: LETTERS TO NEIL ARMSTRONG

The article sent to Carol with this letter from the Vero Beach 32963 *newspaper was entitled "Memories of Neil Armstrong: American Patriot," dated August 30, 2012. Written by James M. Clash, the story featured Clash's recollections of meeting Neil at an annual dinner of The Explorers Club in New York City in 2002. One of the lines from the article read: "In a world where everything is about 'me, me, and me,' he [Neil] was a rare throwback to a time when humility and character counted, when people routinely risked their lives, not to get rich, bloviate or self-aggrandize, but for their country, science and exploration."*

The Bains were good friends of Neil and Carol from the Camargo Club. Like Neil, St. John Bain (1926–2019) was a navy man through and through. Born in Norfolk, Virginia, he enlisted in the navy at age seventeen, becoming the only non–college student selected to participate in the Eddy sonar and radar research program taking place at the Great Lakes Naval Center and Washington Naval Laboratories. Following his military service, St. John earned a BS from MIT, after which he began a thirty-five-year career in Cincinnati with the Formica Corporation. He met Gail "Gay" Seybolt on the courts of the Cincinnati Tennis Club, where he later served as president. He and Gay (who is still living) married in 1956. Playing from the Camargo Club, Bain won not only numerous local tennis championships but also two national doubles championships (with long-time partner Bill Schneebeck) and twice was a finalist; he was also a two-time national senior champion. He once told Neil the following joke, which Neil liked to tell others: Distraught wife: *"You love tennis more than you love me."* Husband: *"That's true. But I love you more than I love golf." Actually, Neil would reverse the order of the joke as he loved golf much more than he loved tennis.*

"BEAUTIFUL CELEBRATION OF AN INCREDIBLE LIFE"

September 13, 2012

Dear Carol,

Neil's public memorial is taking in place in Washington's beautiful National Cathedral as I write, and I know how disappointed my mother-in-law Louise is not to be there. Thank you for inviting Camargo members to Neil's private memorial service. It was a wonderful tribute and beautiful celebration of an incredible life, and those of us there could easily feel how Neil touched so many lives with who he was, and how much he cared despite such remarkable accomplishments. And how blessed to be loved, not only by so many, but by those so close to him. To be loved to the very end of a full and rewarding life—how lucky he was to have you—seems like the best gift of all.

My deepest sympathy for your loss.

Thinking of you,

Nirvani
Nirvani Head
Cincinnati, OH

Nirvani Nyuk Lan Head (b. 1963) is the wife of Jeb Head (b. 1958), the son of Joseph and Louise Head (whose condolence letter dated September 3, 2012, is included in this chapter). Jeb is CEO of Atkins & Pearce, Inc., a company established in Covington, Kentucky, in the year 1817 for the manufacture of braided textiles (cordage and twine, tubing, sleeving, lacing tapes, tie cords, and candlewick). Jeb took over the running of the company in 1986 from his maternal grandfather, Asa Atkins (d. 1989), after working for many years in the banking industry. Jeb and Nirvani have been active in a number of Greater Cincinnati civic and charitable activities, notably ProKids, a nonprofit organization whose mission is to "mobilize our community to break the vicious cycle of child abuse and neglect."

"ONE OF THE GREATEST MEN IN HISTORY"

September 13, 2012

Dear Carol,

I was on a river boat in Europe when I received the shocking news about Neil.

I tried to send you an e-mail to Neil's address, but the message came back as undeliverable.

Neil was one of the greatest men in history—his name will live forever right along with Christopher Columbus and the Wright Brothers. His death is a great loss not only to his relatives and friends, but also to mankind. We shall always miss him.

I am very proud that I knew him and could call him my friend. Even though we did not meet very often in recent years, we kept in touch with frequent e-mail messages. If you have your own e-mail address and if you are interested, I could continue to send you the series of articles entitled "Alex Remembers." I sent those to Neil once every month.

Once again my sincere condolences.

With best regards,

Alex
Sandor (Alex) Kvassay
Wichita, KS

Sandor "Alex" Kvassay is a little-known yet fascinating character with significance in not just American but global aviation. Born in Budapest, Hungary, in 1926, he managed to avoid Hungary's pro-Nazi draft by clerking in the country's defense department. Following World War II, he pursued a law degree while working as a translator and interpreter for the Hungarian Reparations Office, which was formed to administer reparations to the Soviet Union. Secretly he also secretly performed work for the West by photographing encrypted telegrams and codebooks created by the Hungarian Foreign Service, thereby providing information about the rise of communism in his country to the U.S. and British governments. In October 1947 he moved with his young family to Istanbul, where his served as consul general to Turkey. When the Communists took over the Hungarian government and recalled him to Budapest, Kvassay immigrated with his family to the U.S. as a political refugee. Known for his espionage work, he got a job as a translator at the Pentagon. Enlisting in the U.S. Army, he was sent to Fort Riley, Kansas, where he worked as a supply clerk and occasional translator. Upon discharge in late 1951, he returned to Washington, D.C., gaining citizenship and resuming his translating work for the Pentagon. In 1953 he was hired by the Beech Aircraft Corporation in Wichita, where he worked until 1965 as a sales representative to foreign buyers of company aircraft. He left to join the international marketing division of Lear Jet Corporation (bought by Gates Aviation in 1967), eventually becoming its vice president before joining Canadair in 1976. In the late 1970s, Kvassay founded his own jet leasing firm, Management Jets Worldwide, as well as acting as a consultant for other aviation companies until retiring in 1982. His wife, Celia, died in 2010. At ninety-three years of age he now lives in Scottsdale, Arizona. The papers of Sandor Kvassay are preserved in the Wichita State University Special Collections and University Archives. The archival material documents essentially his entire life from 1941 to 2014. His life story merits the writing and publication of a comprehensive biography. His friendship with Neil Armstrong dates from his time working for Gates Learjet.

"IT WAS WE, NOT ME"

September 14, 2012

Dear Carol,

We are writing to express our sorrowful condolences on the passing of Neil, your beloved husband, our friend, and a pioneer hero for the whole world.

I had the pleasure of being the Chief Technology Officer for Eaton Corporation during the time that Neil served on Eaton's Board. It was at several of the Eaton Board events that we enjoyed meeting you and Neil.

Neil was not just a courageous, innovative, and pioneering hero who landed on the moon. But he was the epitome of humility, never taking advantage of his hero status.

In my dealings with Neil, he rarely even talked about the moon landing that he was involved in. And, in those instances when someone asked him about the man on the moon program, he always used the phrase . . . "it was We, not Me."

Neil and I also served together on the board of AIL, and at one early morning breakfast we were talking about solar flares, and for the first and only time, Neil described to me the resultant X-ray emissions from the solar flares and how dangerous they were to the lunar mission. I flew back to Cleveland that night, after dropping Neil off in Cincinnati, feeling accomplished that finally Neil mentioned the lunar mission.

Neil was a brilliant engineer. I fondly recall his technical presentations, on holes and electrons in silicon electron devices, of all things. He always amazed the audience with his knowledge and clear presentation.

Since we both retired, we have exchanged emails, discussing technology and the challenges to engineering universities. My last email to Neil wished him well after his cardiac surgery.

We have been blessed to know this humble man who has inspired us all. We keep you and your family in our prayers.

Stan & Cindy

A RELUCTANT ICON: LETTERS TO NEIL ARMSTRONG

Stan Jaskolski
Dean Emeritus
College of Engineering
Marquette University
Marquette, WI

Neil came to know Dr. Stanley V. Jaskolski (b. 1939) through his service on the board of Eaton Corporation, for which Jaskolski served as chief technology officer. Having earned his bachelor's, master's, and PhD degrees at Marquette, Jaskolski taught in his alma mater's College of Engineering for fifteen years (1967–1982), also chairing its Department of Electrical Engineering and Computer Science (1974–1982) before joining Cleveland-based Eaton (1982–2003), initially as head of its Milwaukee Research Center for thirteen years before moving to Ohio to lead its $350 million-plus R&D program identifying and prototyping high-impact new products for energy management systems. Jaskolski was an extraordinary innovator, with twenty-seven patents in semiconductor technology. His prolific intellect was recognized nationally and internationally, with President Bill Clinton appointing him in 1996 to a six-year term on the National Science Board. Retiring from Eaton in 2003, he returned to Marquette as Opus Dean of the College of Engineering, the first endowed deanship at the university. He retired from the post in 2009. During his six years as dean, he raised more than $120 million for his college and was the driving force behind the planning and fundraising for a game-changing state-of-the-art engineering facility, the Discovery Learning Complex. He also championed recruiting diverse students to pursue engineering, with heavy emphasis on supporting the National Society of Black Engineers.

"THE CATHEDRAL SERVICE WAS BEAUTIFUL AND WORTHY OF NEIL"

September 14, 2012

Carol:

The enclosed is from the Washington Post Op-Ed page of September 13.
 The cathedral service was beautiful and worthy of Neil.
 I thought Gene's [Cernan] tribute was a masterpiece, and I'm glad you picked him to deliver it.

LETTERS FROM A GRIEVING WORLD

Pat and I send all our love to you and yours during this sad time.

Mike
Michael Collins
[Address withheld]

The enclosed op-ed, on page A17, was written by Mike Collins and entitled "The Neil Armstrong I Knew." The concluding paragraph of Collins's tribute read: "Age treated Neil well. As more accolades came his way, he took them in stride. He never showed a trace of arrogance, and he had plenty to be arrogant about. It was refreshing to see him as modest as ever. When my wife, Pat, and I had lunch with Neil and his wife, Carol, this spring, he seemed relaxed, cheerful, contented, happy. I like to remember him that way. He deserved all the good things that came his way. He was the best, and I will miss him terribly."

Michael Collins (b. 1930 in Rome, Italy) was Neil's command module pilot on Apollo 11. Having earned a BS degree at the United States Military Academy at West Point in 1952, he was a fighter pilot and experimental test pilot at the Air Force Flight Test Center, Edwards Air Force Base, California, from 1959 to 1963, logging more than 4,200 hours of flying time. In October 1963 NASA named him to its third group of astronauts. In July 1966 he flew his first mission—the three-day Gemini X mission—during which he set a world altitude record and became the nation's third spacewalker, completing two extravehicular activities (EVAs). Leaving NASA in January 1970, he briefly worked as assistant secretary of state for public affairs before, in April 1971, joining the Smithsonian Institution as the director of the its new National Air and Space Museum, then under construction; he stayed in the job into 1978, two years after NASM's public opening, then moved on to serve as undersecretary of the Smithsonian Institution. In 1980 Collins became vice president of the LTV Aerospace and Defense Company, resigning in 1985 to start his own firm. He is generally considered to be the best writer/author of all the early astronauts; his 1974 book Carrying the Fire *in particular is considered a classic. He was married for over fifty-seven years to the former Patricia Finnegan of Boston, with whom he had three children. Like Neil, Mike Collins has received numerous decorations and awards, including the Presidential Medal of Freedom, Robert J. Collier Trophy, Dr. Robert H. Goddard Memorial Trophy, and Harmon International Trophy.*

A RELUCTANT ICON: LETTERS TO NEIL ARMSTRONG

"SUCH A MISCHIEVOUS SMILE ON HIS FACE"

September 15, 2012

Carol,

I am so sorry for your loss. Neil was such an unbelievable man. One of my favorite memories was when Chris, Jane and I stopped by to say hello. We talked to you for a while and then went downstairs to say hello. You said, "Neil, Chris and Betsy are here" and opened the door to his office. He threw up his arms and had such a mischievous smile on his face & you looked at us like, yep! This mess is his office! We all had a chuckle. He had such an ease and presence about him that I loved. I'm so sorry he's no longer in your life.

I'm sending you lots of love and thoughts your way.

Much love,

Betsy & Chris Vankula
Chicago, IL

Betsy Vankula (b. 1973) serves as Google's head of retail in its Chicago office, a job that has her overseeing digital advertising, marketing, and technology solutions for large retail partners. A BS graduate in communications from Ohio University, she has also led Google's women in leadership program. Her husband, Chris (b. 1971), a graduate of Butler University, also worked for Google but has since created his own company that does management consulting for digital solutions. Betsy is the daughter of one of Carol's best friends from college. Betsy and Carol's son Andrew Knight have been friends since age two.

"HE IS IN HEAVEN LOOKING DOWN ON US"

September 17, 2012

> Some people come in to your life
> and you're never the same AGAIN!

Dear Mrs. Armstrong,

I am so sorry about your very wonderful husband and very special grandfather to my best friend Piper. I know he is in heaven looking down on us and smiling.

Love,

Colleen Cassidy
Darien, Connecticut

Seventeen-year-old Colleen Cassidy is a good friend and schoolmate of Carol's granddaughter Piper Van Wagenen, daughter of Molly Knight Van Wagenen and Brodie Van Wagenen (Brodie is the general manager of the New York Mets major league baseball team). Piper was an extra in the film First Man; *unfortunately, the scene in which she appeared as a waitress did not make it into the final cut. Colleen and Piper are seniors at Darien High School in Connecticut. Colleen's favorite subjects are math and chemistry. She is the captain of the girls' varsity ice hockey team and also the captain of the girls' varsity track and field team. She is EMT certified and works as a volunteer at her local EMS service. She has three older brothers. Her parents are Jacqueline and Kevin (d. 2017) Cassidy.*

"I FEAR THEY DON'T MAKE THEM LIKE THAT ANY MORE"

September 18, 2012

Dear Carol—

Terry and I want to add our expression of sympathy to you on Neil's passing. Like many, we just cannot imagine a world without him in it. How lucky we are to have experienced first hand his wit, intelligence and honor. We are deeply saddened by this, and it leaves a void.

For me, I always associate Neil with my Dad whom I adored. Neil was kind enough to share his memories of Dad and to pay tribute to him at the least expected moments, and I will never forget that. Neil always made me feel proud—of Ohio, of America. He was simply the best, and I fear they don't make them like that any more.

All of us at RTI think of you as part of that family. We all took comfort from your statement about Neil, and my eyelid is getting a lot of overtime with winks to Neil when the view of the moon is clear.

We miss him.

Love,

Ede
Ede Holiday
Washington, DC

Edith E. "Ede" Holiday (b. 1952) is a member of the board of directors of the H.J. Heinz Company, Hess Corporation, White Mountains Insurance Group, Ltd., Canadian National Railway, and RTI International Metals, Inc. She has also served as a trustee for various investment companies in the Franklin Templeton Group of Funds. In 1996 she received the Directors' Choice Award from the National Women's Economic Foundation honoring outstanding women directors for their corporate leadership. With BS and JD degrees from the University of Florida, she was an attorney in private practice before moving full-time to the political arena in the 1980s. In 1982, Holiday became chief legislative advisor and policy aide to U.S. Senator Nicholas F. Brady (R-NJ). In 1984 she was named executive director of President Reagan's Commission on Executive, Legislative, and Judicial Salaries. The following year on until 1988 she worked as chief counsel and national financial and operations director for the George Bush national presidential campaign and political organization. After Bush's election, she became chief spokesperson for the U.S. Treasury Department and soon thereafter the Treasury's general counsel, the first woman to serve in the post. A year later she became cabinet secretary for President Bush; in that role, Holiday was the primary liaison between the White House and all federal agencies. Her husband, Terrence B. "Terry" Adamson (b. 1950), is a former chair of the board of trustees of The Asia Foundation, vice president for global law affairs at Boeing, and executive vice president and chief legal officer of the National Geographic Society. Adamson was also a senior official in the Department of Justice during the Carter Administration, serving as special assistant to the attorney general of the United States and chief spokesperson for the U.S. Department of Justice. Adamson also has served as general counsel, trustee, and executive committee member of the Carter Center in Atlanta since its inception in 1982. A partner of prominent law firms in Atlanta and Washington, D.C., for many years, he was a law clerk to Judge Griffin B. Bell on the U.S. Court of Appeals for the Fifth Circuit. After completing his BA in history and JD with honors at Emory University, Adamson was a Henry Luce Scholar in Japan in 1975–1976. He currently serves on the board of directors of the Henry Luce Foundation.

It is always fascinating to see when the politics of a married couple are so different—in this case, with Ede being such a staunch Republican and Terry being such a strong Democrat, a point that was always made when one of them came up for U.S. Senate confirmation for their nominations to various posts.

"SO HAPPY HE FOUND YOU"

September 24, 2012

Dear Carol,

As our grandparents were neighboring farmers and our mothers were life-long best friends, there wasn't a time I can remember not knowing Neil Armstrong.

He was a very special person and friend. I was so happy he found you with whom to share the remaining years of his life.

Though I hadn't personally seen him recently, I always mailed him a birthday card, just to let him know he was remembered by an old friend. I hope he received this last one.

Sincerely & God Bless,

Carol Crosley Long
Phoenix, AZ

The grandparents of Carol Crosley Long, who lived on a neighboring farm to young Viola Korspeter (Armstrong) outside of Wapakoneta, were Frank B. Crosley (1911–2000) and Georgia Kathleen Hostetler Crosley (d. 1983).

"HE WAS ONE OF A KIND"

September 26, 2012

Dear Mrs. Armstrong,

I recently received notice from Eaton Corporation of Neil's passing along with your address. Please accept my sincere condolences.

As Vice President and Secretary of Eaton prior to my retirement in 1991, I had the honor of being with Neil for several days each month for over a ten-year period while he served as a Director. And I might add he served with quiet distinction.

On occasion we would hold our board meetings at our factories outside the United States followed by golf outings at local clubs. I know Neil enjoyed those trips, knowing he could completely relax in that environment. I hope he saved the photo albums he received after each trip. I think you would enjoy flipping through the colored prints if you haven't already.

One day, probably in the early 80's, after a Board meeting in Cincinnati I was driving Neil to our usual golf outing. He had previously agreed to meet my grandchildren at my home. And we did. They were quite young at the time and were thrilled beyond words when he described his experiences, pointing to the roof of my house explaining he could jump that high on the moon. And then to top it off he kissed my eight-year-old granddaughter as we left.

He was one of a kind. Take care and I do wink when it's appropriate.

Sincerely,

Dick Sadler
Richard T. Sadler
Naples, FL

Neil knew Richard T. Sadler (1926–2017) as corporate secretary and vice president of corporate transportation for the Eaton Corporation. A 1944 graduate of the Staunton Military Academy in Virginia, Sadler served in the U.S. Navy from 1944 to 1946, much of it on active duty on the USS Auger, *a submarine tender, and then with the navy's V-S flight training program. Upon leaving the service he attended Oberlin College and earned a degree in electrical engineering from the University of Wisconsin. Subsequently Sadler obtained his law degree from the Cleveland Marshall Law School (Cleveland State University), after which he worked for several years as an attorney for the U.S. Patent Office in Washington, D.C., before returning to Cleveland and beginning a successful career with Eaton. Like Neil, he enjoyed traveling, playing golf, and building and flying model airplanes.*

"MISSING-MAN FORMATION"

September 27, 2012

Dear Mrs. Armstrong,

On behalf of Strike-Fighter Wing Atlantic, I wanted to send you a note to let you know what an honor it was to participate in and lead the missing-man formation for your husband. The opportunity to commemorate the life of a National hero and former Naval Aviator was a blessing that we were all grateful to be part of!

Sincerely,

Mark W. Weisgerber
Captain, USN
Commander, Strike Fighter Wing Atlantic
Virginia Beach, VA

At the conclusion of the private funeral service held at the Camargo Club on Friday, August 31, 2012, a trio of U.S. Navy F/A-18 jets streaked overhead in a missing man formation in honor of Armstrong's service as a naval aviator. Captain Mark W. Weisgerber (b. 1970) was in command of the formation. A graduate in electrical engineering from Duke University, he joined the U.S. Navy through Duke's Naval ROTC program. Earning his wings of gold as a naval aviator in 1994, he accumulated over 4,000 flight hours with six squadron tours in fleet units flying the F/A-18. He flew during Operations Deliberate Guard, Southern Watch, Iraqi Freedom, and Enduring Freedom, logging more than 200 combat flight hours. Currently Captain Weisgerber is vice commander of the 33rd Fighter Wing at Eglin Air Force Base in Florida. He has earned a number of distinguished awards, including the Legion of Merit, Air Medal with Bronze Star/Combat Distinguishing Device, Navy and Marine Corps Commendation Medals, and Navy Achievement Medals, as well as various campaign and unit awards. He was promoted from commander to captain the week after he flew the missing man formation over Neil's funeral.

"CONSUMMATE TEST PILOT AND AN ENGINEER OF IMPECCABLE JUDGMENT"

October 1, 2012

Dear Carol,

The untimely departure of Neil to the heavens above was very much a shock to Betty Anne and me. In recent years we had been together a number of times which allowed us to become even closer as friends than when we had been colleagues in the spaceflight world. We enjoyed our time with him so much and it was such a pleasure to remember our times together as aeronautical engineers and as a test flight junky. Although we had spent untold hours discussing what we thought was required to land on the Moon, the later times we spent mostly talking about our airplane experiences at NACA. He and Betty Anne seemed to always hit it off when we met at the functions we attended which was quite a thrill for her.

In the world of manned spaceflight Neil and I always seemed to be on the same page together as we tried to decipher the many problems we had in going to the Moon and back. We both recognized we had the same objectives in doing it well and doing it safely. It was a wonderful experience to attack what we both considered the difficult tasks of making it happen. He was indeed a consummate test pilot and an engineer of impeccable judgment.

In recent times we came together in thinking about the future of spaceflight and how we thought it should be approached. Neil's willingness to spend his precious time speaking out to the Congress and the media showed his commitment to the future of our country. He will be missed in many ways but none as great as his advice to all of us as we tried to convey the message of what should be done as we try to choose the next steps in the human exploration of space.

You should also know how we at JSC [Johnson Space Center] and NASA felt about the choosing of the astronaut to first step on the Moon. We made a very deliberate decision that Neil should be the person we wanted to be our national representative. We met, recognized and discussed how this person would be remembered in history and we wanted to make certain that choice was not done without very careful consideration. As I look back on that decision, all I can say is thank God we did what we did.

Neil Armstrong was a great American, a wonderful engineer, a fabulous test pilot, a teacher and above all a great human being.

Neil had a wonderful life and I am sure you were a very great part of that. People like him are extremely rare and I am thankful for the time we spent together. He will be remembered for his many accomplishments and for the significant part he played in opening the path to space exploration.

Betty Anne and I offer our sincere condolence to you and your family. We know he will be terribly missed.

Sincerely,

Chris Kraft
Christopher Kraft
Houston, TX

Christopher Columbus Kraft Jr. (1924–2019) is best known as the legendary original flight director for all of NASA's early space missions, including all of the Mercury flights and many of the Gemini missions. Born in Phoebus, Virginia, adjacent to the National Advisory Committee for Aeronautics' (NACA's) Langley Memorial Aeronautical Laboratory (later NASA Langley Research Center), it was perhaps natural for Kraft to become a devoted student of flight, earning his degree in aeronautical engineering from Virginia Polytechnic Institute (Virginia Tech) in December 1944. Joining the research staff at NACA Langley as a flight engineer, he was assigned to the lab's Flight Research Division under the leadership of Robert Gilruth and Hewitt Phillips. Under their leadership, Kraft contributed to many critical flight-test programs, including evaluations of the flying qualities of aircraft and free-fall model tests to measure transonic and supersonic aerodynamics. With the establishment of NASA in October 1958, he moved with Gilruth into the Space Task Group, which was responsible for putting together and then operating Project Mercury, the United States' first project to put a man in space. Besides serving as the first NASA flight director, he directed the design and implementation of the Mission Control Center at the Manned Spacecraft Center (later Johnson Space Center) in Houston, from which all of NASA's manned space flights would be conducted. In 1970 Kraft became deputy director of the Manned Spacecraft Center and, in 1972, its director, succeeding Gilruth. He retired from NASA in 1982 but stayed active in the aircraft industry through frequent consulting for various corporations. His 2001 autobiography, Flight: My Life in Mission Control, *is considered a classic. The recipient of numerous awards and distinctions, in 2011 Johnson Space Center renamed its Mission Control Center the Christopher C. Kraft Jr. Mission Control Center in his honor.*

A RELUCTANT ICON: LETTERS TO NEIL ARMSTRONG

"RARE TRAIT BUT SOMETHING NEIL HAD IN ABUNDANCE"

October 2, 2012

Dear Carol,

Pam and I are sorry that we did not connect with you at the time of Neil's service in Washington to express our sadness at his passing. But I would suppose that in that week or so you had enough expression of condolence to last a lifetime. It was a most moving service, especially the Bishop's homily, and you carried it all off with great dignity.

    As you know my time with Neil was before he became famous. He was just one of us. It worked out that both of our nomadic fathers landed in Upper Sandusky for the same three years, which proved to be a very formative period for both Neil and I and also Kotcho. In scouts, at the bakery and in classes the three of us were very close. My mother says that the phrase she heard most from me during those years was "me, Neil and Kotcho."

    Neil and I reconnected during his first two years at Purdue. We played in the Marching Band together, had some common classes and met for meals and a beer from time to time. But our prime goal, particularly the first year, was to survive amid the competition from older, mature returning vets.

    Pam and I attended the launch and got reconnected with the family but Neil was otherwise occupied at the time and so it was some years afterward that we met again.

    And then of course we had a number of meetings in Aspen with the two of you and Kotcho. I often recall a very revealing incident from those visits. We were at the Crystal Palace for a show and dinner and after dinner we were going through the process of splitting up the bill with credit cards. At the time American Express was advertising on the basis of card members who were famous and I asked Neil if he had been approached. He reluctantly said he had been. I asked Neil the amount common for these endorsements which I recall was $50,000. You then piped up with some verve "the man can't be bought." That says a lot about Neil and his reaction to fame and I think explains why he is so revered beyond just walking on the moon. All of us, the whole nation, yearn for heroes and leaders who are like that. It's a rare trait but something Neil had in abundance.

Pam and I are now just settled in to a cottage in a very nice retirement community in Exeter NH and still trying to sell our larger house in the country. We also have a summer cottage on the coast of Maine (Cushing) where we spend the summer with family sailing and boating and eating lobster. In Cushing we have become acquainted with another summer resident, Nellie Taft of Boston, who came from Cincinnati and I believe knows you. Several years ago her brother was visiting her and we met at a party and exchanged news about Neil.

Pam and I wish the best for you personally as you adjust to your new circumstances. It was a joy to be with you both in our visits to Aspen and to observe your close relationship and how happy Neil was with his marriage. You obviously were a help and support to him as he continued his unique relationship with fame.

We would be delighted to reconnect up here if your travels take you this way.

Pam joins me with warm wishes.

Bud
John Blackford
Exeter, NH

John "Bud" Blackford (b. 1930) was one of Neil Armstrong's best boyhood friends. In the early 1940s they became close as grade schoolers when their families lived in Upper Sandusky, Ohio, a small town some sixty miles south of Toledo located near the headwaters of the Sandusky River (thus named "Upper"). In the company of another local boy of their same age, Kotcho Solacoff, Bud and Neil became very active members of the Wolf Patrol, part of the town's Boy Scout Troop 25. Theirs was an unforgettable friendship—Bud, Kotcho, and Neil. Even after the Armstrong family moved to Wapakoneta in 1944 and the Blackfords to Hopkinton, New Hampshire (part of Greater Concord), in 1956, they stayed in touch. (Kotcho Solacoff would remain in Upper Sandusky, where as an adult he would set up a highly respected family medical practice.) As for Bud Blackford, he would help to start up the HMC Corporation in Contoocook, New Hampshire, which manufactured machinery for the lumber industry. In retirement, Blackford became a management counselor to CEOs of small manufacturing companies.

A RELUCTANT ICON: LETTERS TO NEIL ARMSTRONG

Neil never forgot either of them. In July 1968 both Blackford and Solacoff and their wives were invited to watch the Apollo 11 launch from the VIP tent.

Today at age eighty-nine, Bud Blackford lives in a retirement community in Exeter with his wife, Pamela.

"NEIL'S SELFLESS ACTS WERE SO KIND"

September 14, 2013

Mrs. Carol Armstrong and Family
Cincinnati, OH

Dear Mrs. Armstrong and Family,

I write to express my deepest condolences for your loss of Neil. I intentionally delayed writing this letter, as my experience is that the loss that is initially felt is met with an outpouring of sympathy and support, but that the long-term grief that is experienced after the death of a loved one is not met with that which is needed down the road.

I write to tell you that your Neil was, indeed, an important role model to me. He did his job well—perhaps one of the all-time understatements. He lived a life of humility and service, quietly doing good works and deeds. His father and mother were kind to me, and were encouraging to me as examples of people who did good things and helped people quietly, without fanfare or personal glory.

And his parents were proud of their children, including Neil. Viola spoke of the many things that happened to them during Neil's career, and she shared some of the funny and interesting stories, and you could tell how amazed she was at the specter of the historical significance of the NASA program and her son's role in it. And her pride, even through her humility, showed in her beaming face.

Neil's selfless acts were so kind. His signing autographs on prints of our courthouse so that we could raise money for Auglaize County Crippled Children's Society and the Auglaize Historical Society for a project recognizing the 100th anniversary of the courthouse building in 1994, helped raise thousands for both charities. We quietly, without publicity, were able to help many children and their families here who were facing difficult times and challenging disabilities, and we gave local people a nice opportunity to obtain a remembrance of Neil and our courthouse. He spoke to me after the project concluded, and shared some thoughts about our efforts, and I really appreciated his taking the time to call me. He wanted a couple of the prints for his sons, expressing how much he liked the painting. He was just so nice.

I doubt that you remember me, Carol, but Mrs. Sandy Brading introduced us at the airport in New Knoxville while Neil and Charlie [Brading] were sitting together as Neil talked about the aerial maneuvers being performed and talked about his memories of airshows and aviation. My late wife and my two older children were sitting on a blanket watching the airshow and listening to Neil, and I ran into Sandy as I was walking across the lawn there.

I lost my wife, Diane, to inflammatory breast cancer in 2006 after a two-year battle with the disease. She was 51 when she died, and my children were 25, 20, and 9 years old. On July 29, 2012, I lost my father, Joe, who died at age 86. Dad was my best man at my wedding in 1975, and was one of my heroes. In my lifetime, I have had very few men I considered my heroes . . . real role models. My dad, an uncle who fought on Omaha Beach, my pastor J. C. Herbert . . . and Neil Armstrong.

I write this to let you know that my experience is that grief is a terrible, but necessary, thing that just keeps on coming. I experience it daily in one form or another, and I have come to accept it as necessary in keeping life in perspective. Missing a loved one takes many forms. Sometimes it is filled with laughter and fond remembrances, and sometimes with tears. Other times it is filled with anger, or depression. My son is now a doctor in Seattle, my older daughter is in medical school at Ohio State, and my youngest is now coming to her junior year in high school. We all miss their mother, and each experience grief differently.

So please know that the terrible grief that you have experienced, and will continue to experience, is something that you are not alone in. And know that your beloved Neil will continue to be a role model for so many . . . not only in "heroic acts" (his life certainly was filled with courageous acts and his ultimate "cool under pressure") but especially for his quiet, humble and effective example. My grandmother always urged me to be a "do-er," meaning to be someone who got things done rather than one who talked about doing things but never got anything done. Nail was just that . . . a "do-er."

In 2012, I offered to give Neil a tour of the renovated courthouse if he wanted to see how the renovation of our courthouse had turned out after nearly two years of being closed for renovation. That offer stands for you or any member of your family. If any of you would like to see the renovation, including the restored statue of Lady Justice that the Auglaize County Historical Society arranged for in the 1994 celebration I referenced above, please call, write or email my office and I will be glad to show them to you.

And, again, please accept my sincere condolences for your loss. Neil is missed by so many, me among them.

Very truly yours,

Frederick D. Pepple
Wapakoneta, OH

In 1994 Armstrong did, in fact, sign 250 prints of a painting of the Auglaize County Courthouse to raise money for the charities indicated in Frederick Pepple's letter. As a condition for signing them, Neil required that the prints be sold "only in person, in that county," so that the prints would go not to collectors (for resale at higher prices) but to local residents in Ohio. To make it as difficult as possible to forge his signature, Neil, on every print, put a faint pencil mark near his signature, a mark only visible when the print was held to the light in a certain way. For many years there was, according to Robert Pearlman of collectspace.com, "no record of any of these prints being sold." That situation changed after Neil's death in August 2012, and changed even more dramatically during the fiftieth anniversary of Apollo 11 in 2019, when a number of the Auglaize courthouse prints were sold to collectors.

Frederick David Pepple (b. 1954) has served for many years as the general division judge on the Auglaize County Court of Common Pleas in Wapakoneta. After earning his undergraduate degree from Bowling Green State University in 1975 and his JD from Ohio Northern University in 1978, he worked as a general practice attorney in Wapakoneta and Waynesfield, becoming the prosecuting attorney for Auglaize County from 1981 to 1986 before election into his initial term as judge.

APPENDIX

Secretaries, Assistants, and Administrative Aides for Neil Armstrong, 1969–2012

There was no way for Neil Armstrong to keep up with the heavy volume of mail he received over the years following Apollo 11 without considerable—weekly if not daily—assistance from secretaries or administrative aides. While he was still working for NASA, the agency did what it could to help Neil manage his mail by assigning a person or two from Public Affairs to open his mail, read each letter, and make a judgment as to whether it was a letter Neil would want, or need, to see. The assistant might also answer the letter, following guidance Neil had provided, always making sure the reply would satisfy him. Anything the assistant was not sure about, she most certainly ran before him, taking the side of caution, to make absolutely sure the mail was being handled exactly how Neil wanted. Naturally, as an assistant gained experience working on Neil's mail, she became more confident in her decisions and did not need to bother him as much in the triage of his correspondence.

Another thing the assistants did, particularly the assistants at NASA Public Affairs, was to keep stats on Neil's incoming and outgoing mail—that is, how many letters, cards, and so forth were received each month and how many replies the office sent on his behalf.

The following table provides the names of all of the women who helped Neil with his mail at NASA, at the University of Cincinnati, and through private arrangement with him once he went to work for himself after leaving the university in 1979. Of course, most people did not know when exactly Neil left NASA, arrived at Cincinnati, or left Cincinnati. Few knew

APPENDIX

Years	Name	Organization
1969–1971	Shirley B. Weber	NASA, Office of Public Affairs
1971–1972	Geneva B. Barnes	NASA, Office of Public Affairs
1972–1973	Fern Lee Pickens	NASA, Office of Public Affairs
1971–1973	Ruta Bankovskis	University of Cincinnati
1974–1975	Luanna J. Fisher	University of Cincinnati
1976–1979	Elaine E. Moore	University of Cincinnati
1980–2003	Vivian White	Private (Lebanon office)
2003–2012	Holly McVey	Private

his working address, and even fewer knew his home address. So, letters kept arriving at NASA after he left, with the same being true after he left the university. For a short period after he left both institutions, his previous assistants continued to do what they could to help. After leaving the University of Cincinnati, Neil had to arrange for the administration of his mail and soon realized that handling his mail on his own was an impossible burden. In February of 1980 he rented a small office and post office box in Lebanon, Ohio, a little town north of Cincinnati, where the Armstrong family had moved to a farm after Neil resigned from NASA in 1971, and hired an administrative assistant—Vivian White—who managed his correspondence for more than twenty years. While he was at NASA, letters requiring translation were handled by government employees and contractors. At the University of Cincinnati, such letters were handled as best as could be from within the university community. The same was true when letters requiring translation arrived in the Lebanon post office for Vivian White to handle, as Vivian herself could not translate the letters.

While Neil was still with NASA and during his first couple of years at the University of Cincinnati, he also relied on these assistants to help him with his scheduling, bookings, and travel itineraries. Starting in 1974 he referred requests for nonacademic appearances—and there were a lot of them—to Mr. Thomas Stix, Stix and Gude, 30 Rockefeller Plaza, New York, New York 10020.

We know a great deal about how Vivian White handled her assignment with Armstrong, as she was interviewed at length for *First Man*; and from Vivian we also know a lot about Neil's policy for signing autographs.

APPENDIX

Vivian White worked full-time for Neil for about ten years; after that she "cut back" to four-and-a-half days a week. According to Vivian, for the first twelve to fifteen years she worked for him, Armstrong would sign anything he was asked to sign, except a first-day cover. Then in about 1993 he discovered that his autographs were being sold over the Internet and that many of the signatures were forgeries. So he just quit signing. Still letters arrived in Lebanon saying, "I know Mr. Armstrong doesn't sign anymore, but would you ask him to make an exception for me?" After 1993, form letters under Vivian's signature went out in answer to 99 percent of the requests. In the few instances that Armstrong accepted the invitation, he composed and signed a personal letter. If he chose to answer someone's technical question, according to White, he would "write out his answer, I'd type it up and then put underneath it, 'Mr. Armstrong asked me to give you the following information,' and I signed it. We never answered personal questions. They were just too much an invasion of privacy." In Vivian's filing system, they would go into "File 11," the wastebasket.

It has been difficult to find biographical information about Armstrong's administrative aides, with the exception of Geneva B. Barnes and Vivian White. Thanks to a lengthy oral history interview conducted with Geneva Barnes on March 26, 1999, by historian Dr. Glen Swanson at NASA Headquarters as part of the NASA Johnson Space Center Oral History Project, we know a great deal about her. We also know quite a bit about Vivian White as she was interviewed at length as part of the research for *First Man*. Profiles of both Geneva and Vivian follow.

PROFILE OF GENEVA B. BARNES

Geneva B. Barnes—known to her friends as Gennie—was born on June 29, 1933, in Tahlequah, Oklahoma, so she was roughly three years younger than Neil.

During her senior year in high school in Tahlequah, Gennie was encouraged to enter into government service by her business administration teacher, who urged her to apply for a job with the Navy Department in Washington, D.C., knowing that a civilian recruiting officer would be visiting their school prior to Gennie's graduation, looking for talented

APPENDIX

stenographers. Along with her two best friends, she took the civil service test and received an appointment with the Office of Naval Material in Washington. It was the beginning of what turned out to be a forty-one-year career of federal service.

Gennie worked at the Office of Naval Material for four years, then accepted a position at the Pentagon in the Office of the Judge Advocate General (JAG). From there she moved to a position assisting the director of the Washington Regional Office of the Post Office Department (now the U.S. Postal Service). She was still working at the Post Office Department on February 20, 1962, the day Mercury astronaut John Glenn made America's first orbital flight in space. She left her office and stood in the rain watching Glenn in the company of Vice President Lyndon B. Johnson riding in a parade down Pennsylvania Avenue celebrating the historic mission. Afterward, she went back into her office, telephoned the NASA personnel office, and asked if it was hiring secretaries. Invited to apply, Gennie filled out an application with NASA the next morning and was, in her words, "on the payroll by 10 o'clock."

Her first job with NASA was as a secretary in the Office of Programs at NASA Headquarters. Then in 1963 she moved to the Office of Public Affairs, where she worked for the next eight years, handling many of the behind-the-scenes arrangements involved with NASA special events, including White House ceremonies and astronaut award ceremonies and appearances. During the Apollo missions, she assisted in protocol activities at Kennedy Space Center for four of the flights, including Apollo 11. One of the grandest and most personally rewarding events of her life was being part of the select group of support staff who accompanied the Apollo 11 astronauts and their wives on their Giant Step Presidential Goodwill Tour around the world from September 29 to November 5, 1969, visiting twenty-three countries in thirty-eight days. It was no vacation, however, as Gennie had worked hard with members of the State Department and President Nixon's White House staff setting up every detail of the project, including preparation of briefing materials and schedules.

As revealed in Armstrong's letters, Gennie served as Neil's public affairs assistant during his stint as deputy associate administrator for aeronautics in NASA's Office of Aeronautics and Space Technology. When Neil left NASA, she stayed on at OART until 1980, when she for a brief period worked as a management analyst for NASA's Office of Management Operations. But Gennie loved the astronauts and soon returned to a post as coordinator of astronaut appearances for the Office of Public Affairs. When she retired from NASA in 1994, she was handling the astronauts' appearances when they traveled internationally. After her retirement, Gennie spent many hours doing volunteer work at the White House in the e-mail section of the Presidential Mail Office.

In retirement Gennie lived in Capitol Heights, Maryland, just outside of the District of Columbia, due east of the nation's capitol.

PROFILE OF VIVIAN WHITE

Vivian White was born in 1921 to parents Archie and Lucille (Currey) Tartt in Kings Mills, Ohio, twenty-five miles northeast of Cincinnati. After graduating from Kings Mills High School in 1939, she began a secretarial position with Lou Romohr, a realtor in nearby Lebanon, Ohio, who later became the town's mayor. Vivian worked for Rohmer for twenty-eight years, during which time she earned her own real estate license, the first women in the history of Warren County, Ohio, to do so. In 1980, Lebanon's chamber of commerce honored her as Woman of the Year.

That same year Vivian went to work for Neil Armstrong, an assignment she handled with skill and dedication for the next twenty-three years, until 2003, from an office Neil rented in downtown Lebanon.

Vivian had two daughters, Lois and Margie, three grandchildren, four great-grandchildren, and two great-great-grandchildren, as well as several step-grandchildren and -great-grandchildren. She had one sibling, a brother named Archie Eugene Tartt. Besides working for Neil, Vivian enjoyed collecting antiques and playing bridge. She was active in the Town and Country Garden Club and Lebanon Council of Garden Club, organizations in which she held several leadership positions. She also was a long-time member of the Lebanon Business and Professional Women's Club.

Vivian White died on November 27, 2017, at the age of ninety-six.

NOTES

1. "Collection Focus," Barron Hilton Flight and Space Exploration Archives, http://collections.lib.purdue.edu/flight-and-space/info.php.
2. Wernher von Braun quoted in George W. Cornell, "Space Expert Sees New Awareness of God," *Cleveland Plain Dealer*, July 19, 1969; Pope Paul VI quoted in *Baltimore Sun*, July 21, 1969, A4.
3. Reverend Herman Weber quoted in "St. Paul United Church of Christ Pastor Offers Prayer," *Wapakoneta Daily News*, July 16, 1969; Sermon by Reverend Charles Sloca, Fairfield, Iowa, "Vision via Television," copy in Viola Engel Armstrong Papers.
4. Dan L. Thrapp [Los Angeles Times/Washington Post Service], "Moon Walk to Shift Man-God Views," *Dayton Journal Herald*, July 19, 1969.
5. Ming Zhen Shakya, "A Nobel Prize, Lunar Communion, the Beatitudes, and a Song of David's," accessed October 10, 2019, https://zbohy.zatma.org/Dharma/zbohy/Literature/essays/mzs/beatitudes.html.
6. Dudley Schuler to James R. Hansen, August 15, 2002, transcript p. 7. Schuler was a classmate of Armstrong's at Blume High School in Wapakoneta.
7. John Grover Crites quoted in "Astronaut's Home Town Swept by 'Moon Craze,'" Syracuse (NY) *Post Standard*, July 4, 1969, and "Moon Was a Dream to Shy Armstrong," *Dayton Journal Herald*, July 11, 1969; Crites quoted in "Neil Armstrong—All American Boy," *The Blade Sunday Magazine, TB*, December 5, 1965; Eugene Kranz to James R. Hansen, Dickinson, Texas, December 10, 2002, transcript p. 27; Charles Friedlander to James R. Hansen, San Diego, California, April 8, 2003, transcript p. 27.

NOTES

8. "Face the Nation as Broadcast over the CBS Television Network and the CBS Radio Network, Sunday, August 17, 1969—11:30 a.m.–12:30 p.m.; Origination: Houston, Texas; Guests: Crew of Apollo 11; Reporters: Walter Cronkite, CBS News, David Schoumacher, CBS News, Howard Benedict, Associated Press," CBS transcript, p. 24; Dean Armstrong to James R. Hansen, November 14, 2002. This comment was not recorded and thus is not part of the official transcript.
9. "Space Window," Washington National Cathedral, accessed October 23, 2019, https://cathedral.org/what-to-see/exterior/space-window/.
10. "Space Window at the Washington National Cathedral," *Atlas Obscura*, accessed October 10, 2019, https://www.atlasobscura.com/places/space-window-at-the-washington-national-cathedral.
11. James R. Hansen, *First Man: The Life of Neil A. Armstrong* (New York: Simon and Schuster). See the 2005 and 2012 editions, pp. 426–32, and 2018 edition, pp. 221–22.
12. Mission Operations Branch, Flight Crew Support Division, NASA Johnson Space Center, Houston, Texas, vol. 1, pp. 6-33–6-37, https://www.hq.nasa.gov/alsj/a11/a11tecdbrf.html.
13. J. Allen Hynek quoted in Curtis Fuller, *Proceedings of the First International UFO Congress* (New York: Warner Books, 1980), 156–57.
14. Jimmy Carter quoted in Robert Sheaffer, *UFO Sightings: The Evidence* (Prometheus Books, 1998), 20–21. The incident was also related in Howard Raines, "Carter Once Saw a UFO on 'Very Sober Occasion,'" *Atlanta Constitution*, Sept. 14, 1973, 1D.
15. Carol Armstrong to James R. Hansen, June 3, 2004, transcript p. 12.
16. James Smith to James R. Hansen, July 17, 2003, transcript p. 14.
17. Neil A. Armstrong to James R. Hansen, June 2, 2004, transcript p. 8. The cassette tape recordings of all of my interviews with Armstrong, as well as all of the verbatim transcripts of those interviews, are preserved in the Neil A. Armstrong papers collection in the Purdue University Archives and Special Collections, West Lafayette, Indiana. All interviews with Armstrong cited in these notes took place at his home in suburban Cincinnati.
18. Sheila Burke, "Nashville Filmmaker Confronts Former Astronaut Buzz Aldrin," February 9, 2002, accessed at Tennessean.com.
19. Edgar Mitchell quoted in online article published in October 2006 by BC Skeptics, British Columbia Society for Skeptical Inquiry, accessed October 30, 2019, at https://web.archive.org/web/20061002085955/http://www.bcskeptics.info/resources/skeptopaedia/index.cgi?key=sibrel%2C%20bart.html.

NOTES

20. Armstrong's response can be seen and heard at the following video link: https://www.youtube.com/watch?v=vGqZj5dci_M&list=PLOeF2Q2PeojU80Mo0qnEdRmEcKzUcWlcF.
21. Armstrong to Hansen, September 22, 2003, transcript p. 1.
22. James C. Fletcher quoted in "Neil Armstrong to Leave NASA," NASA Flight Research Center *Express* 14 (August 28, 1971): 2.
23. Armstrong to Hansen, June 2, 2004, transcript p. 10.
24. Armstrong to Hansen, September 22, 2003, transcript p. 13.
25. "About the National Space Society," accessed October 10, 2019, https://space.nss.org/about-national-space-society/#targetText=The%20National%20Space%20Society%20(NSS,creation%20of%20a%20spacefaring%20civilization.&targetText=The%20society%20also%20publishes%20Ad,most%20important%20developments%20in%20space.
26. "Offshore Technology Conference (OTC)," accessed October 10, 2019, https://www.energy.gov/eere/wind/events/offshore-technology-conference-otc.
27. Buzz Aldrin, "Commentary: Let's Aim for Mars," CNN, June 23, 2009, accessed October 10, 2019, http://edition.cnn.com/2009/TECH/space/06/23/aldrin.mars/index.html.
28. Neil A. Armstrong email to James R. Hansen, June 27, 2005.
29. Konstantine Solacoff letter to James R. Hansen, May 16, 2003, p. 5.
30. Viola Armstrong to Dora Jean Hamblin [*Life* magazine], tape 1a, pp. 5–6, copy of transcript provided to me by June Armstrong Hoffman, Neil's sister, during preparation of *First Man*. I also had access to the original tape recordings in the Time Life archives in New York City.
31. Armstrong to Hansen, September 22, 2003, transcript p. 30.
32. Ibid.
33. Ibid., p. 31.
34. Ibid., pp. 33–34.
35. Charles Lindbergh quoted in A. Scott Berg, *Lindbergh* (New York: G. P. Putnam's Sons, 1998), 537.
36. Armstrong to Hansen, June 4, 2003, transcript pp. 22–24.
37. "Professor Norman Drummond Delivers the Mountbatten Lecture 2016 at Edinburgh University," *ww100scotland* (blog), June 13, 2016, http://ww100scotland.com/blog/professor-norman-drummond-delivers-the-mountbatten-lecture-2016-at-edinburgh-university/.
38. "Bronze Medal," Royal Scottish Geographical Society, accessed October 28, 2019, https://www.rsgs.org/bronze-medal.
39. Crystal Chesters, "Moon Landing 50 Years On: Why Astronaut Neil Armstrong

Calls Scotland Home," *The Scotsman*, last updated July 19, 2019, https://www.scotsman.com/heritage/moon-landing-50-years-on-why-astronaut-neil-armstrong-calls-scotland-home-1-4963833.

40. "The Lone Sailor Award," United States Navy Memorial, accessed October 28, 2019, https://www.navymemorial.org/lone-sailor-awards-dinner.

41. Indeed, the entire world grieved upon hearing the news of Neil Armstrong's death. The grief can be seen in the many headlines and newspaper, internet, and other media stories that appeared in country after country around the globe in the days and weeks that followed his death on August 25, 2012. This final chapter of *A Reluctant Icon* does not actually document the global reaction, however; in that sense, the title of the chapter is a little misleading. What the chapter actually presents are messages of condolence sent to Neil's widow, Carol Armstrong, in the weeks following Neil's death, as well as some letters of sympathy sent to the entire Armstrong family. What is most striking about these letters is what they represent about Neil's remarkable circle of friends during his lifetime, especially during its final decades. Virtually every message of condolence published in this chapter comes from an extraordinary individual with a brilliant career of his or her own—in science and technology, in business, in government and politics, in the military, or in some other arena of professional endeavor. The identity of the letter writers shows how being the First Man on the Moon—nay, just being Neil Armstrong, the thoughtful, highly intelligent, hard-working, productive, active, flourishing, enormously respected and greatly loved man—brought forward a life of tremendous opportunity, diverse experiences, long-lasting friendships, and high-powered networks of professional connections and association. It isn't that Neil did not have many friends who were just ordinary, everyday "little people"; he had many of them and treated them just as respectfully and faithfully as the host of VIPs and notables that were bound to come into his life as a very famous man and American hero. But the people who wrote condolence letters to Neil's widow were naturally going to be those who knew Neil and Carol best, as their friends—and, for that matter, who simply knew the home address of Neil and Carol Armstrong in suburban Cincinnati so that their letter of condolence could reach its appropriate destination.

42. According to certain family members with whom I have spoken, the text of the Armstrong family's message to the public concerning Neil's death was drafted by Neil's good friend and Cincinnati neighbor Joseph W. Hagin Jr., a veteran Washington, D.C., political aide who, along with his father, Joseph Hagin Sr., was a good friend of the Armstrongs. Before its release for publication, the

letter was offered to family members for editing, with a number of changes made by Neil's youngest son Mark Armstrong. Some members of the family thought that the final version of the message ran inappropriately long—given that Neil was a fan of brevity. Some also did not like the "winking" part, though they kept that sentiment to themselves. As Neil's biographer, I do prefer to think of him whenever I see an airplane flying overhead, as his original and everlasting passion was for aircraft; however, I admit that there are not many times when I look at the Moon that I don't think about Neil as the Apollo 11 commander who first piloted a flying machine down to the lunar surface.

43. "Presidential Proclamation—Death of Neil Armstrong," The White House Office of the Press Secretary, August 27, 2012, accessed October 29, 2019, https://obamawhitehouse.archives.gov/the-press-office/2012/08/27/presidential-proclamation-death-neil-armstrong.

44. See Robert Z. Pearlman, "First Moonwalker Neil Armstrong Mourned at Washington's National Cathedral," Space.com, September 13, 2012, accessed October 17, 2019, https://www.space.com/17585-neil-armstrong-mourned-national-cathedral-memorial.html.

45. See "Neil Armstrong, Who Stood on Moon's Sea of Tranquility, to Be Buried at Sea," Collectspace.com, September 6, 2012, accessed October 17, 2019, http://www.collectspace.com/news/news-090612b.html.

46. Shane Scott and Sarah Kliff, "Neil Armstrong's Death and a Stormy, Secret $6 Million Settlement," *New York Times*, July 23, 2019, accessed October 17, 2019, https://www.nytimes.com/2019/07/23/us/neil-armstrong-wrongful-death-settlement.html. A follow-up article on Armstrong's death, the settlement with the hospital, and the sale of Armstrong memorabilia by his two sons appeared in the *New York Times* on July 27, 2019, entitled "'Would Dad Approve?' Neil Armstrong's Heirs Divide Over Lucrative Legacy" and written by Scott Shane, Sarah Kliff, and Susanne Craig, can be found at https://www.nytimes.com/2019/07/27/us/neil-armstrong-heirs.html (accessed October 17, 2019). As Armstrong's authorized biographer, I was interviewed for and quoted in both articles.

47. I acquired access to all ninety-three pages of the documents that had been sent anonymously to the various media; the narrative of this chapter introduction is based on those documents.

48. See Robert Z. Pearlman, "Neil Armstrong's Family, NASA Remember First Moonwalker," Space.com, August 31, 2012, accessed October 17, 2019, https://www.space.com/17415-neil-armstrong-memorial-service-nasa-family

.html. I attended the private funeral, which took place on the grounds of the Camargo Club in the Cincinnati suburb of Indian Hill, a club to which Neil and Carol had belonged for many years.

49. "Unexpected Sources of Inspiration," USI, accessed January 10, 2020, https://www.usievents.com/en/.

50. https://www.nydailynews.com/news/crime/paul-goresh-shot-photo-john-lennon-killer-dies-article-1.3760884, accessed January 10, 2020.

ABOUT THE EDITOR

James R. Hansen is professor emeritus of history at Auburn University in Alabama. An expert in aerospace history and the history of science and technology, Hansen has published a dozen books and numerous articles covering a wide variety of topics, including the early days of aviation, the history of aerospace engineering, NASA, the Moon landings, the Space Shuttle program, and China's role in space. In 1995 NASA nominated his book *Spaceflight Revolution* for a Pulitzer Prize, the only time NASA has ever made such a nomination. His book *First Man*, which is the only authorized biography of Neil Armstrong, twice spent three weeks as a *New York Times* Best Seller—in 2005 and again in 2018—and garnered a number of major book awards. Translations of *First Man* have been published in more than twenty languages. A Universal Studios film adaptation of the book hit the silver screen in October 2018, with Academy Award winner (*La La Land*) Damien Chazelle directing the film and actor Ryan Gosling starring as Armstrong. Hansen served as coproducer for the film, which won 28 major awards, including an Oscar and a Golden Globe, and received 190 other major nominations.

Hansen began his career in aerospace history while serving in the early 1980s as historian-in-residence at NASA Langley Research Center in Hampton, Virginia. In his first book while working at Langley, *Engineer in Charge* (1985), he uncovered the story of the segregated group of African American women—including the late Katherine Johnson (1918–2020), a Presidential Medal of Freedom recipient in 2015—who worked as

ABOUT THE EDITOR

mathematicians and data processors for the government laboratory, later made famous in the celebrated 2016 film *Hidden Figures*.

Over the years Hansen has served on a number of important advisory boards and panels, including the Research Advisory Board for the National Air and Space Museum, Editorial Advisory Board for the Smithsonian Institution Press, and Advisory Board for the Archives of Aerospace Exploration at Virginia Polytechnic Institute and State University. He also is a past vice president of the Virginia Air and Space Museum in Hampton, Virginia. For the past ten years he has served on the National Air and Space Museum Trophy Selection Board. His experience as an academic and public speaker has been wide-ranging both topically and geographically; he frequently serves as keynote speaker, panelist, and lecturer on a wide variety of topics in the history of science and technology.

Hansen has received a number of prestigious awards for his scholarly contributions to the history of flight, notably the Eugene M. Emme Astronautical Literature Award of the American Astronautical Society, Gardner-Lasser Aerospace History Award of the American Institute of Aeronautics and Astronautics, and Robert H. Goddard Prize of the National Space Club.

A native of Fort Wayne, Indiana, Hansen graduated summa cum laude and with high honors from Indiana University. He earned his master's and PhD at The Ohio State University in 1976 and 1981, respectively. Hansen taught history at Auburn University from 1986 until his retirement in May 2017. He was chair of the Auburn history department for four years and director of the Honors College for six years. Both his teaching and his scholarship received numerous awards from Auburn, including induction into the College of Liberal Arts Teaching Hall of Fame. Students who earned graduate degrees under Hansen's direction have held positions at the University of Central Florida, National Aeronautics and Space Administration, National Air and Space Museum, U.S. Air Force Academy, U.S. Air Force Air War College, National War College, Pentagon, and American Society of Mechanical Engineers.

Hansen has been married to Margaret Miller Hansen, also a Fort Wayne native, since 1976. They reside in Auburn, Alabama, and have two children and four grandchildren.

CPSIA information can be obtained
at www.ICGtesting.com
Printed in the USA
LVHW021813040820
662394LV00014B/248

9 781557 539694